The Meaning(s) of Life

Life

A Human's Guide to the Biology of Souls

by **M**

Contents

1 | Introduction

What Hath Biology Wrought

> *There are only two ways to live your life. One is as though nothing is a miracle. The other is as though everything is a miracle.*
>
> – Albert Einstein

It is a scene repeated in baseball parks all across America. A renowned power hitter hits a home run, altering the course of a game. As he crosses home plate, he looks up, pointing one or both fingers to the sky. I live in the Boston area, so I often see David Ortiz do this. You come to expect it. It's part of his ritual. A similar meme has found its way into other American sports. Even if you are unfamiliar with the rules of the particular game, or are in no way religious, you implicitly understand the gesture. The athlete is publicly attributing some part of his success to a higher power. He points to the sky because that is where this higher power lives, a variation on calling out the contribution of a teammate by pointing in his or her direction.

But where, exactly, is David Ortiz pointing? If you trace the trajectory from his finger skyward, up from Fenway Park, through the Earth's atmosphere and on into space, it just keeps going, endlessly, to no place in particular in a vast and possibly infinite universe. Since the Earth rotates on its own axis, as well as orbiting the Sun, each time he points, he is most likely pointing to a different non-place in the universe. If his Japanese counterpart were to make a similar gesture at the same time (though they probably wouldn't both be playing at the same time), he and David Ortiz would be pointing in opposite directions. If one is pointing "up," the other must be pointing "down."

A few thousand years ago, this conceptual anomaly would not have occurred to (most) humans. We thought we lived on a more or less flat surface. We weren't terribly concerned about how far it extended, or whether it had any ultimate edges. The sky was universally up, the Earth universally down. The place where the higher being resides could be some fixed place in particular. You could reliably point to it. Now, of course, we have a more nuanced perspective. We know we really live on a sphere, one of uncountably many that are spread out over vast distances in a vast universe. We've seen pictures of our blue sphere from space, and other spaceborn pictures of the more exotic stellar phenomena that surround us. Even those who believe in a Heaven would not be comfortable trying to locate it somewhere in this modern universe. It's not so much that we were wrong about the correct location of Heaven, as that the notion of Heaven being in a spatiotemporal location no longer makes sense. The two worldviews are incommensurable.

But our old worldview was *approximately* true, at least the geometry parts. When you live on the surface of a very large sphere, it looks like a plane. Gravity gives you a universal proxy for up and down. Most of us will never personally experience anything that exceeds this convenient approximation. So we live most of our lives on a virtually flat Earth, just like our ancient ancestors. We comfortably speak of up and down, of sunrises and sunsets. We know which teammate David Ortiz is calling out. When we think of the real geometry of the cosmos, we go through a conceptual gestalt shift. We snap, discontinuously, to a different perspective. (It's rare for this to happen during a baseball game.) Because of the approximate truth of the flat world, some humans – scientists, transcontinental airline pilots, astronauts – can make this switch smoothly. There are analogs in the new worldview for most concepts from the old one. Heaven is not one of these.

Our distant ancestors also got around with a fundamentally different concept of life than we have today. And as with cosmology, we still carry many of the vestiges of that view around with us, causing us to go through similar kinds of biological gestalt shifts. There are just a lot more of them than in cosmology, and there are very few points of smooth transition between the old biology and the new biology. This is because our ancestral theory of life, the theory of souls, is not an approximation of the modern biology reality. It is something else entirely, a fundamentally different notion of what makes living things lively. Souls are units of disembodied agency, something that gets added to physical bodies to make them living, and accounts for the amazing, purposeful things they do. For the longest time, even after scientists had figured out the mechanics of the universe, they still couldn't nail the fundamental mechanics of living things. So souls hung around as the default, everyman's explanation of life well into the 20th century. They are in our laws, our literature, our social customs. This sometimes puts us in the conceptually awkward position of trying to locate souls in bodies (in the brain?), or locate where and when they enter bodies (in the uterus? – but only in mammals?), or when they leave bodies (what about comas and vegetative states?). No one really wants to go there, because next you'll have to figure out how much souls weigh, and the dynamics of their energy transfer with the body, and other such silly-sounding things.

So why do we go there? Because for many people, the gestalt shift back to molecular biology has the unsettling consequence of *reducing* a person to a body. Oh my, that can't be right! (Or, we hope it isn't right ...) What about love, and empathy, and beauty, and kindness, and irony, and happiness? There must be something more to us than just molecules! Well, yes. But the whole point of science is to understand complex things by reducing them to the predictable interactions of simpler, more fundamental things. You know you've succeeded when you can use the newly discovered underlying principles to reliably reproduce the behavior of the complex things, or to alter their behavior, or to make new versions of them.

Humans appear to have no problem at all with this kind of scientific reduction of non-living things. It's cool. And this has nothing to do with understanding. Most of us have no idea how utterly strange it gets when you reduce things to the atomic level (let alone the quantum level). When David Ortiz strikes the ball with his bat, we hear a very solid crack, and his hands feel the violent concussion of the collision (much more so if he misses the sweet spot). This is the very essence of solid and hard in folk-physics. But the atoms that make up the bat and ball are mostly empty space. If an atom's orbiting electrons formed a sphere the size of Fenway Park, its nucleus would be about the size of a small insect sitting in shallow center field. With all of that space between the particles that make up the ball and bat, the chance of any two of them colliding is pretty slim. So why doesn't the bat pass right through the ball? Electromagnetic forces. Think: like poles of two magnets repelling each other – that squishy, continuous, invisible force-field kind of thing. The ball atoms and the bat atoms retain their separate spatial integrities by collectively attracting their own members and collectively repelling the other bunch. The very solid cracks and thuds of the baseball world are not properties of their constituent atoms. These properties emerge as atoms are aggregated into molecules, and bats, and balls, and baseball fans to hear the compression waves, and baseball players to feel the recoil.

Emergence and reduction are simply opposite directions of the same relationship. It is very common for complex systems to have emergent properties that are not just the sum of their parts. That's how this very interesting universe of ours works. It makes no sense to say that hard and solid are just illusions because they have an atomic reduction to things that aren't hard and solid. Instead, we say that *this is what hard and solid are.* So why all of the angst about reducing love to neurons and hormones, or free will to our inability to predict what the molecular causes of our behavior are? If you are distressed by what's lost on the way down, why not marvel instead about what's gained on the way up? Why do we insist there must be something more to love, when we don't seem to care about there being something more to the hardness of bats and balls? Well, because we aren't either of those things. But we *are* living things. We have a ringside seat, on the inside as it were, that lets us feel what it's like to be a human. Lives are us. It's personal. We have no vested interest in "the meaning of stars," or "the meaning of mountains," or "the meaning of houses." But "The Meaning of Life!" For that we use capital letters. It is compelling. We want to know!

Now let's consider this meaning-of-life thing. What exactly is that? We know it as one of those "big" concepts, something very profound. You have to climb to the top of a mountain and ask an obscure bearded man in a white gown about it. It's some sort of secret, not obvious to any of us. We are always seeking it, never finding it.

We inherit a version of this concept from our predecessors in which 'meaning' means 'purpose.' What is the purpose of life? *The* purpose, the canonical one, the one true one. This carries with it the implicit notion that there is some sort of distinguished creator, or orchestrator, of life whose purpose this is. Someone or something – a God, a pantheon of Gods, Mother Nature, the Force, the Masters of the Universe – was responsible for human existence, and did not orchestrate all of this fortuitously. He/she/it/they had something particular in mind. It behooves us to figure out what this is, so we can align our own goals with it instead of working at cross-purposes. This book will be of no help to you in such a quest. You will need to find someone who claims to have privileged access to this orchestrator, then figure out why they and not one of their many competitors is the designated messenger. There are plenty of books on that already.

One of the concepts that science shed on its way to growing up was the notion that there is an inherent purpose for every class of things in the universe. This was not necessarily a religious idea. It did not involve creators. We owe it to Aristotle who just thought it made for a compelling sort of metaphysics. *Final causes* they were called. Like the location of Heaven, this is one of those ideas that just didn't fit in as the science advanced. Ordinary causes turned out to be good enough. We still have purpose. It is all around us. Many living things have purposes – goals, things they try to accomplish, motivations that explain their behavior. Other things, like mountains, don't. Life itself is one of these other things. It doesn't have a purpose – just like red, or beauty, or thunderstorms don't have a purpose. All of these things have causes, though. Even purposes have causes. And these causes turn out to be the best resource we have to understand the nature of things.

Notice that traditional questions surrounding the meaning of life – "Why are we here?" "Why are we like this?" "Could it have been some other way?" "Why is it so hard to find happiness?" "Is there someone out there for me?" – are much easier to address with causes than purposes. For life to have a purpose, it requires a cause anyway. And not just any cause, but a particular kind of cause: a creator whose purpose it is. It's hard enough to discover anything about such a cause, but even harder to get any insight into the unknowable mind of this creator. What was he/she/it/they thinking? If you just let science do its work on the actual causes of life and human nature, you stand to learn something useful about why we are here and why we are this way. This is that kind of book.

In the large, the meaning of life is more than a definition of life (though we have one of those). It is a complex collection of meanings, and relationships to other concepts. We have laws and customs and mores and therapies and social relationships and interventions and lab protocols that all depend on life being a certain way, lasting a certain amount of time, being feasible in certain environments, being amenable (or not) to alteration or creation. This is common among our more fundamental concepts. They get their meanings as large wholes. As the constituent relationships change over time, the core meaning drifts a little. Generally these changes can be managed by making little adjustments here and there. Conceptual change happens slowly. But every once in a while, scientists discover that they have been so fundamentally wrong about some key concept that there is no way to gracefully adjust the rest of the whole. It becomes incoherent. Something once thought to be a fundamental truth must be jettisoned in order to bring the whole set of concepts back to consistency.

This kind of thing happened in the early 20th century, for example, when Einstein introduced principles of relativity into physics. He discovered that the geometry of our universe has some surprising features that do not match the geometry we had been assuming for the previous 2000 years. That geometry, first defined by the Greek mathematician Euclid, is based on five postulates that had been taken to be self-evident – part of the very *meaning* of geometry some would say. But two of those meanings could not both be squared with how the universe is actually laid out: that the shortest distance between two points is a straight line, and that parallel lines are everywhere equidistant. Since nothing can travel faster than the speed of light, the shortest distance between any two points in the universe is the path traced by a ray of light between them. Gravitational fields, however, can "bend" those paths, so two such "straight lines" that are equidistant in some places, may converge or diverge in others. One of these assumptions had to go, so the fifth postulate, the parallel postulate, was dropped. It turns out we live in a non-Euclidean world.

Recently, this same kind of thing has been happening to our traditional concept of life, only more so. We formed our original cluster of meanings and relationships concerning life based on big human and mammalian life – the sorts of creatures humans are likely to encounter in their normal environments. Mammals give birth live to a (mostly) fully-formed, though smaller, version of the parent, so we missed a lot of subtleties about embryonic development and metamorphosis that link single cells to whole animals. We also had no idea about the natural origin of all of this stuff, so we missed the very fundamental progression from micro life to macro life and how the evolving ecologies of life at these two different scales make our planet work. For big creatures to directly beget big creatures, you need some fairly miraculous transitional principle. So we solved this, along with the problem of the begetting of the first of the big creatures, with the blunt force instrument of the soul. Life was a rolling miracle of disembodied agency that touches down in successive generations of bodies.

In the last 150 years or so, we figured out the principles that allow complex life to evolve from simpler life, bringing enormous unification and elegance to the processes of biology, and allowing us to fill in the gaps from one big creature to the next with small, incremental, non-miraculous steps at the micro level. In the last 60 years or so, we figured out the molecular machinery that makes all of this stuff work below the threshold of life. In the last 20 years or so, we have discovered that human, and even animal and plant, life is by far the least representative variety of life on the present Earth, and that there are viable living things in every nook and cranny of what we used to classify as uninhabitable spaces. In the last 7 years or so, we have discovered (via induced pluripotent stem cells) that life can be made to run backward – and it still works! Every year, every quarter, every month we find more ways to repair life, to re-program life, to synthesize life, to extend life, to bend it, twist it, and pull it into forms once unimaginable.

In many ways, life has turned out to be fundamentally different than we expected. Our once simple, unified meaning of life is being shattered into many, sometimes competing, concepts. This is what happens near the beginning of conceptual revolutions. *The* meaning of life is beginning to lose its coherence. We, or our successors, may eventually be able to put this all back together into one concept, but right here, right now, we are more like all the King's horses and all the King's men sifting through the shards of Humpty Dumpty. Some of the things we've discovered recently still fit the original contours, allowing us to understand how once mysterious life processes at the macro level actually work, often in surprising and ingenious ways that are more awe inspiring than the miracles they replace. Some of these discoveries render old notions of life completely incoherent. And some will require us to re-imagine old concepts of life in yet-to-be-determined ways.

That is the gist of the '(s)' in the title of this book. We are presently faced with many meanings of life. We are fortunate (or unfortunate, depending on your point of view) to be living at a time when these discoveries are first coming to light. It has never been harder to grasp the significance of being human than in this newly nuanced world. But we also have more degrees of freedom than we ever imagined. So what's a human to do?

One approach is to look the other way, to pretend that none of this is really happening. Most of the present humans on this planet do not believe in evolution. They are a little behind, about 150 years or so. This is not just a matter of under-exposure to biology. Non-belief among industrialized nations is at its highest in the United States where an estimated 97 million people reject the evolutionary explanation of their own origin. In some southern states, there are government officials who believe it is still possible to roll back the clock of biology education to a simpler time, and a simpler concept of life. As if this would somehow make it go away.

Another approach, for those who fear that the coming biological reductions will strip us of our humanity, is to hope that this will not happen too soon, or perhaps that some mysteries will never be solved, or that the science fiction scenarios about creating and altering and indefinitely extending life will remain fiction. These folks are a little behind the news as well. It's already happening, in laboratories all over the world.

Still another approach is to thank your lucky stars that you were born during this golden age of biology, and get to witness these developments firsthand. It is just plain fascinating to see how our human nature emerges from molecules, and what that might tell us about why we are the way we are.

This book is very much in the spirit of the last approach. I have arranged the chapters in a roughly small life to large life order, reflecting the emergence point of view. Some of the biology is drawn from what has been known (to scientists, anyway) for some time, but is useful to recapitulate as organizing principles for what comes later. Some of it reflects recent changes in the biological perspective that are not widely known yet outside of science. And some of it is really new or still very speculative, the kind of things you wouldn't normally hear about unless you closely follow this stuff. My aim is not to teach biology, but to select snippets of biology that are now interestingly at odds with the way we are accustomed to thinking about life. It is also not a prescriptive book. It will not teach you how these new perspectives will lead you to a better life. It is mostly a useless book, intended to make you wonder about the meaning of life in new ways – for no reason in particular.

2 | What Is Life?

*In three words I can sum up everything I've learned about life:
it goes on.*
     – Robert Frost

Well, we know it when we see it. Or at least we think we do.
But because we personally experience very little of the variety of life
that nature has to offer, we often get it wrong. Our most general folk-
taxonomy of things in the universe has three categories: animal,
vegetable, mineral. We use these categories in guessing games as a
way to quickly narrow down the space of things we have to choose
from. No doubt, our ancient ancestors had a similar classification
scheme: two kinds of life, animal and plant, and everything else.
Animals are generally lively, characterized by autonomous
movement. Plants don't move, but they grow. Everything else doesn't
move (by itself) and doesn't grow.

But this scheme breaks down pretty quickly when you stress it.
Rivers and wind both move and grow. OK, living things have to move
and/or grow *of their own accord.* They have to generate these
behaviors internally, not just get pushed and pulled around by their
environments. But storms and wildfires generate their own growth
internally. They even reproduce smaller versions of themselves.
Then there are the hybrid cases. Bone, wood, leather, fur, textiles are
all mineral things that were once animal and vegetable things. Death
is not a firm boundary, though, because many of these things were in
their current mineral state when they were still parts of animals and
vegetables. The outer edges of living things – the calcium in bones,
the bark in trees, the last layer of epidermis and hair in animals, the
external filaments of cotton plants – are connected continuously to
still living, or barely living, versions of themselves as you move
inward.

Corals certainly look like minerals, but they are actually animals, still very much alive. At the lower end of the animal and plant kingdoms, the visible behavior can be very confusing. Some animals, such as sponges and sea anemones, are sessile – permanently anchored to the sea floor. They don't move. Some of their plant counterparts, like algae, swim. Finally, consider seeds and spores. Plant seeds, such as acorns, are desiccated, embryo versions of the parents that have ceased metabolism indefinitely until the proper environmental cues wake them up to start "living" again. They are not dead. Spores are the bacterial equivalent. The whole organism closes up shop and shuts down inside a protective coating to last for however long it takes to find suitable living conditions again.

In order for science to classify corals and sponges as animals, and algae as plants, and bacteria as something else altogether, it needs to discount things like movement and growth, and identify what is truly essential to all living things. Science likes to be as general as possible, so we would like to characterize what is essential to life anywhere in the universe, just as we attempt to characterize all planets or all stars, not just the ones in our vicinity. We think life ought to be a general phenomenon, a natural kind. The problem is that we don't have any examples of life anywhere else in the universe, not even in our own solar system. Consequently, there has always been a suspicion that life is a one-off kind of thing. Maybe it is so improbable that it only happened once – right here. Every instance of life that we have ever discovered has the same carbon-based biochemistry, the same amino acid and DNA/RNA polymers, and one of two cell types. So it is tempting to bake this into the definition of life and settle for "life as we know it." As a practical matter, when scientists search for "habitable" planets outside our solar system and scan them for signs of life, they look for atmospheric signatures of our version of the biochemistry – water vapor, oxygen, ozone, and carbon dioxide – because this is one of the few effects of life you can observe about a planet from many light years away. Since we don't know any competing biochemistries, we wouldn't know what other effects to look for.

We don't mean to be so parochial, so back here on Earth, where we can get up close and personal with it, we attempt to characterize life by the *functions* that this particular biochemistry makes possible – what characteristics do all instances of life, large and small, have in common. We already know of a few alternative nucleotides and amino acids that can be substituted for the natural ones without breaking organisms, so we allow for the possibility that other chemistries and other molecular organizing principles might implement these functions in some other manner. The general consensus among scientists gives us three modern functional concepts to replace the three ancient concepts – not two kinds of life and one kind of non-life, but three features that anything we will call life must possess: 1) metabolism, 2) a semi-permeable boundary (such as a cell) that isolates it from the rest of its environment, and 3) heritable reproduction. The second criterion is a curious one because it seems like it might be implied by the other two – you can't get a successful combination of metabolism and reproduction without some sort of environmental boundary. Or at least we think so. We haven't seen it done any other way. So the boundary criterion could possibly turn out to be like the parallel postulate.

If you made it this far, you may be wondering: hey, what's with all the biochemistry? I thought this was a book about the meaning of life, in the big sense – you know, about human nature and souls and love and free choice and stuff like that. Well, hang in there. We'll get to that. We are recapitulating life in the order that it originally came about. All of that more inspiring stuff is ultimately composed out of the (possibly) less inspiring stuff we are about to cover. This chapter lays the groundwork for life that will be used by all of the rest of the chapters. Feel free to skip ahead, though you may find yourself coming back, because there is a lot of backward reference. On the other hand, you may be inspired by all of this molecular stuff as well. In that case, welcome aboard.

Imagine that the original Intelligent Designer of life on Earth was busy working on other planets, and chose you as the intelligent assistant to work out the details for our planet. Before you could inquire about how to go about this, he/she/it/they was out the door and off to some other part of the universe. So how would you go about this monumental task? Try following this guide.

How to Make a Metabolism

Most of us are familiar with the concept of metabolism, if at all, in the context of our own bodies. If you have a fast metabolism, your body burns food at a quicker rate, so you don't gain weight as easily as someone with a slower metabolism. When you consume alcohol, you remain in an inebriated state, and thus vulnerable to a drunk driving citation, until your body metabolizes the excess alcohol in your blood. It is a kind of burning, an extraction of energy from fuel that your body uses to power its operations. Like a fire, it is an inherently unstable phenomenon, an ongoing chemical reaction that depends on a steady source of fuel. Without more fuel, the fire eventually goes out. The reaction stops. The life in the body dies.

One of the remarkable things about this "fire of life" is that it is one of the few processes in the universe that creates complex order out of chaos. The rules of the universe are rigged in favor of chaos and randomness. We call the measure of this randomness *entropy*. On average, most natural processes are headed in that general direction. Life goes the other way. To see how life manages to pull off this anti-entropic trick, we have to delve into a little bit of thermodynamics (but not much).

Entropy

Thermodynamics is a branch of physics that is concerned with energy flow. It had its origin in the 19th century study of steam engines, and was originally aimed at the engineering problem of designing more efficient engines. It has yielded four laws (numbered zero through three) that describe general relationships that hold across processes in nature that involve energy transfer. The zeroth law defines the notion of temperature. The first says that energy is neither created nor destroyed; it just changes forms. The second says that whenever energy changes forms, the effect on the universe as a whole is always a net decrease in the quality of total energy – its capacity to do work. And the third says that nothing can reach a temperature of absolute zero. The second law gets all of the attention because it posits a fundamental asymmetry in nature. Natural processes have a one-way direction. The capacity for the fixed amount of energy in the universe to create order is gradually winding down. Each time this capacity is used to create some order, a little bit of it leaks out as heat, never to be recovered. There are no natural processes that can result in a net improvement in energy quality. There aren't even any that leave it the same. This doesn't prevent nature from improving the quality of energy *locally* – from creating *local* order – it just requires a compensating destruction of order somewhere else. Plus a little extra.

The upshot for steam engines is that you can't design a perfectly efficient one. Only some of the energy consumed can be converted to useful work. Some fraction will always escape as dissipated heat. This also ruled out the possibility of a perpetual motion machine. How we got from steam engines to the universe at large has to do with the notion of system boundaries. Since in the natural world everything seems to be constantly affecting everything else, spreading energy around willy-nilly, it is impossible to conduct any controlled experiments to measure the net effects of energy flow. The parts of steam engines, like boilers and piston cylinders, provided a nice, tidy, artificial environment where a rigid container isolated the processes occurring in a volume of water or steam. Temperatures and pressures could be controlled rigorously with valves and heat sources. As a result, the thermodynamic principles were defined relative to *closed* systems – systems that could be sealed off from interaction with their surrounding environment so that one could observe what happens spontaneously, once all of the initial conditions are set. The result concerning nature's tendency to seek equilibrium at the state of greatest entropy (least order) is a statement about what will happen in such a closed system. Once the system is sealed off from the rest of the world at a certain pressure and temperature, whatever natural processes can occur in this environment will continue to occur, each time losing a little bit of order, until no more processes are possible. This is the state of equilibrium. And since every transaction strictly decreases order, the final state represents maximal disorder. If the container starts with an unevenly heated volume of water, for instance, the heat will gradually spread out until it is uniform at every location. The initial temperature gradient cannot last, and it certainly can't get more graded. If the container starts as a volume of water with a drop of indigo ink deposited in one spot, the ink will randomly spread out until it is uniformly distributed and the water is a uniform pale blue color. If the container starts with various concentrations of chemicals, they will continue to react with each other, catalyzing intermediate products until all of the reactions that can happen have happened. Each reaction will consume a little bit of free energy and sacrifice some of it to heat. When there is no longer enough free energy to power any more reactions, we will be at equilibrium with a final set of products that can no longer react.

The irony of developing the entropy concept in the context of closed systems is that it only really applies to the real world in either very local, artificial systems that can be engineered to be isolated from their surroundings, or to the universe as a whole, where the system is closed by the boundary principles of cosmology. It is the ultimate generalization from observed laboratory facts. At almost every level of organization in between, systems, if they can be said to exist at all, are open systems. Their boundaries are permeable, and they exchange energy and matter with their surroundings. In this context, the entropy law doesn't apply to their internal goings-on because the spontaneous internal processes are constantly being influenced by forces from the outside. So there is no imperative to drive toward equilibrium. We are surrounded by thermodynamically open systems that run in this out-of-equilibrium state, in the form of nature's organisms and our own human-built machines, so we normally don't observe the creeping entropy that is building up on the balance sheet of the universe as a whole. It looks very orderly to us. There was a time when cosmologists thought that the entropy principle implied that the universe would ultimately wind down to a final "heat death," but now they are not so sure. But even if that were the ultimate fate of our universe, the small energy degradation implied by each order-producing transaction of our many open systems will take a very long time to add up. It's a very slow leak.

Biostorms

It is difficult for nature to create thermodynamically open systems like metabolisms, and even harder to keep them running for any significant duration. It's a matter of getting a continuous chain of reactions to keep happening in the same local area. If the system is too open, the reactions will just dissipate into the surrounding environment, losing their local coherence. If the system is too closed, the reactions will eventually reach equilibrium, exhausting all the possibilities. Wildfires are an example of both. The chain reactions radiate out from the original center in all directions while the center reaches an expanding equilibrium as the fuel is exhausted.

Firefighters bring them to an end by effectively converting them to closed systems with backfires that consume all of the fuel at the periphery.

To get an open chain reaction to continue occurring locally, you need to somehow get a feedback cycle started – the chain must form a circle: A causes B causes C causes A. This is hard to accomplish on a flat surface, a la forest fires, but easier in atmospheres and oceans. Perhaps the best-known examples of these naturally occurring feedback cycles are the circular storms: cyclones, anticyclones, and tornadoes. These arise in the atmosphere when temperature and pressure gradients happen to get together in the right order. Once the feedback cycle develops, the system perpetuates itself by drawing energy from the surrounding environment. When these favorable feeding characteristics are exhausted, the system loses energy and coherence, eventually dissolving into equilibrium. Although we tend to think of these things as rather short-lived phenomena, they can persist indefinitely in the right environments. Jupiter's big red spot is an anticyclone that has been going on continuously for centuries. The storm is about 2-3 times the size of the Earth and completes its cycle about once every six Earth days. Our own planet exhibits an impressive variety of these self-sustaining cycles at the global scale that have been operating continuously for millions, and even billions, of years. We have the cycle of volcanism and plate tectonics that recycles our crust, the global weather system that perpetuates dependable trade winds, the cycling patterns of ocean currents that keep our climates temperate. We have a much greater variety of these kinds of dynamic systems than our rocky planet neighbors, and this is thought to be one of the reasons why we had inherently better initial conditions for the origin of life. But it is difficult sorting out just what caused what, because some of the dynamic cycles of our planet are themselves caused by the presence of life, including the carbon cycle and the presence of significant oxygen in our atmosphere.

Strictly speaking, we don't call these self-perpetuating cycles of energy transfer *metabolisms* except when they occur inside living organisms. And here they are ubiquitous – trillions upon trillions of tiny biostorms, self-catalyzing hurricanes of molecular reactions among proteins, nucleotides and other small and large molecular factors that occur inside individual cells. This constant swirl of reactions breaks things down to free up energy, and uses it to build up new things that power the outward focused behaviors of cells. These in turn sum up, in multicellular creatures, to the complex movements of whole animals through networks of sensor, neuron, and muscle cells. Life is more than just metabolism, but metabolism is what supplies the continuous molecular motion that distinguishes life from other collections of inert stuff. We suspect that Earth's pre-biotic molecules exhibited more variety than those of Mercury, Venus, or Mars, and that this is why we were able to develop the critical variety necessary to spawn molecular biology. We infer this from afar, due to the greater apparent variety and complexity of our weather. And the most distinguishing feature in this regard appears to be our oceans of liquid water. We have them. Our neighbors don't. Our oceans provide not only an ideal starting medium for carbon-based biochemistry, but they also serve to regulate global temperature and to spawn lots of local variety in weather, which in turn creates a variety of local environments, increasing the chances that some of them will spawn and support emerging biostorms.

The storms themselves are a kind of order. They may look like chaos, but they persist because the swirling reactions form a stable pattern – a circular pattern. This is a *dynamic* order. The system constantly changes but the overall pattern of change doesn't. From a sufficient distance, it looks like a solid thing (a la Jupiter's spot). *Static* orders, like buildings, degrade their share of energy in the original construction, but then stop. Dynamic orders keep degrading energy continuously just to be an order at all. There just aren't enough of these kinds of anti-entropic processes occurring in the wild to make any significant dent in the order deficit of the universe. The neat trick about life is that it managed to bottle up some early, successful biostorms, from their very open and transient chemical configurations, inside containers at the Goldilocks point of permeability – not too open, not too closed, but just right. This incubation of biostorms inside bioreactors converted them from transient events to relatively permanent things. And as we will see in the next section, the flexible properties of these containers enabled them to subdivide as the cycles inside grew larger. This put life into the order-manufacturing business. Big time.

How to Contain a Metabolism

The bioreactor does two things to keep metabolisms going. It is just closed enough to keep the reactants in close proximity, and to prevent their escape, and it is just open enough to allow food molecules in and waste molecules out. So it needs to be a special kind of container that is tailored to the roles of the molecular species involved, letting the right ones in and keeping the wrong ones out. In the lab, molecular reactions are typically contained in a glass beaker. The reaction can continue because the beaker prevents the diffusing reactants from drifting away. They find their partners and catalysts because they keep bouncing off the glass and returning to the fray. The right food molecules are let in either by tubes and valves, or just by pouring them in the opening at the top, exploiting gravity to prevent escape through the same opening.

Nature's Beaker

At the large, multicellular level of animals, such as ourselves, the beaker glass that incubates our metabolism is our skin and gut lining. Topologically, we are like a tube, with a large opening at one end (mouth and nose) to let oxygen and food in, and a smaller one at the other end (anus) to let the waste products out. Our bioreactor is the space between the inner and outer surfaces of the tube. That doesn't amount to much space in a typical tube, so instead imagine a large cylindrical tank with a much slimmer tube running down its center. You still have two surfaces: the outer surface of the cylinder itself, including top and bottom, and the inner surface of the internal pipe connecting the top to the bottom. In macroscopic animals, the outer surface (skin) is mostly impermeable to environmental molecules and other organisms (such as bacteria). It is designed to keep things out in general. The inner surface (gut lining) is semi-permeable to these smaller things and designed to let the right things in. But since the intake opening to the whole organism is very large, lots of things that can't get through the skin (such as the wrong bacteria) flow right in through the mouth and nose. So we have an additional layer of security standing guard at the inner surface in the form of our immune system. We share this general design with all multicelled animals right down to microscopic worms.

At the single-cell end of the spectrum, the shape of the container approximates a sphere. There is only one surface exposed to the outer environment, so everything that gets in and out has to pass through the cell membrane, and thus has to be small-molecular in size. Single-celled creatures have no mouth and gut. So if they need nutrients that naturally occur as small parts of much larger clumps of things, they have to pre-digest their food by secreting digestive enzymes into the environment in the vicinity of the clump and wait for it to break down. We perform this same task inside our guts. Some single-celled creatures, such as amoebae, employ a hybrid version of this (called *phagocytosis*) by reforming their shape to surround big food, then closing off the open end of the shape to form an internal vesicle containing the food – not unlike how we open and close our mouths to ingest a whole bite of steak. Digestive enzymes then break it down on the inside.

The final act of metabolism though, the great cycles of molecular reactions that power the core business of life, always takes place inside individual cells. The semi-permeable cell membrane that keeps the right molecules in and the wrong molecules out appears to be nature's universal solution for conserving the reactions of life in a beaker. We don't find life organized into any other format. The cell is life's atom. There is no life at a unit of less than one cell. And every higher form of life is ultimately composed of cells. This singular solution to the system boundary problem for life suggests that it must be pretty optimal.

There are other ways to contain metabolism-like things and keep them going that do not result in living things. If we take an abstract, functional view of metabolism – a circular system of causes and effects that maintains out-of-equilibrium stability by consuming inputs and producing outputs – then certain features of ecosystems could be considered metabolisms. Wherever food chains are able to perpetuate themselves due to internal cycles that ultimately depend on the Sun for the continuous supply of new input energy, we have the functional realization of a metabolism. Green plants eat light from the Sun, carbon dioxide from the air, and nitrogen from the bacteria in their roots to make sugars and respire oxygen. Herbivores eat the sugars and oxygen. Carnivores eat the herbivores and oxygen. Both respire carbon dioxide. Bacteria and fungi eventually eat the dead plants and animals and the cycle continues.

The system boundaries that keep the reactant organisms in sufficient proximity are geographical – ponds or forests or mountain ranges. These kinds of fixed boundaries, though, don't permit the contained system to move around as a unit or faithfully reproduce itself. The cell membrane does, and so its contained metabolisms become units of life.

Modern cells come with a variety of additional container structures. Plants have outer cell walls made primarily of cellulose that jointly support the integrity of their plant's shape. Many bacteria have relatively rigid walls made of peptidoglycan in between two lipid bilayer membranes. Fungi have outer walls of chitin, the same material that forms the exoskeletons of crustaceans and insects. Algae have outer walls made of several different substances. Animals, both the single and multicellular variety, have no cell walls. But all cells have the same kind of flexible membrane, and its function in each is to provide the selectively permeable boundary between the inner metabolism and the outer environment.

The Cell Membrane

Now, suppose you set out to design such a cell membrane to solve the metabolism containment problem – some kind of surface whose permeability is right at the Goldilocks point for life. It would also have to be strong, able to hold its spherical shape, but fairly flexible, able to shape-shift and grow as the contents grow. You would also need some process for shaping it into spheres to begin with, and for gracefully budding off new spheres at reproduction time. This would be a non-trivial task even for an intelligent designer. But suppose there just happened to be some naturally occurring molecule hanging around the environment whose chemical properties give you all of this for free. Well lucky you, lucky us, lucky for life, there is!

The magic molecule is called a *lipid*. There are several types of these, but what they have in common is a head portion that is naturally attracted to water, and a tail portion that naturally repels water. Because of these contrasting affinities, when a bunch of them get together in water, they quickly assemble themselves into a two-dimensional sheet composed of pairs of molecules with the two tails facing each other and the two heads facing outward in opposite directions. This is called a *lipid bilayer*. Both sides of the surface of the sheet attract water (because of the heads), but the layer of tails sandwiched in between repels water. So water-attractive molecules, like proteins, carbohydrates, nucleic acids and amino acids – the basic building blocks of life – can get right up next to either side of the surface but will not pass through it because of the hydrophobic (water fearing) center layer.

The individual lipid molecules in the bilayer are not attached to each other by chemical bonds, they just prefer to associate with each other in this uniform pattern. Because of this, the sheet is somewhat flexible and can be bent into shapes, including spheres, and can grow and shrink by movement of lipid molecules into and out of the formation. In fact, the biophysics of this molecule will cause it to self-assemble into two additional forms: the *micelle*, a tightly packed sphere whose outer surface is all water-loving heads and whose inner volume is just the water fearing tails; and the *vesicle*, a larger sphere whose surface is itself a bilayer with an inner volume of water. This last shape, of course, is the basic structure of the cell membrane. It is nature's universal solution for biological containment tasks. In addition to using it to house their metabolisms, cells use the formation of internal lipid vesicles to transport biostuff from point A to point B inside the cell. Cells that eat big things via phagocytosis bring them inside by surrounding them with vesicles that they "bud in" from their outer surfaces. Many internal structures of complex cells, such as the nucleus that holds the DNA, and the mitochondria that produce the energy, are housed in these spherical bilayers, though it is likely that both of these "organelles" were once cells themselves (see chapter 4).

Micelles will naturally merge with vesicles to make their surfaces larger. And vesicles have some natural biophysical affinities for merging together and subdividing, similar to what happens with soap bubbles. Nature prefers spheres because they are often the simplest solution to least energy problems, so forming spheres, budding into two spheres, and merging into one sphere, come with the territory. The pharmaceutical industry has learned to exploit these properties in artificial lipid vesicles to make drug delivery vehicles.

How the first membrane came to encapsulate the first metabolism – whether one preceded the other – will always be a matter of speculation because the soft structure of cells leaves no fossils and because metabolic events are transient. What scientists try to do instead is to simulate possible scenarios in the lab to determine what *could* have happened. We know that the modern version of the cell membrane is probably quite evolved from the original proto-cell (it's had 3.5 billion years to experiment with improvements), so researchers try to simulate what might have happened with the more primitive chemistry of pre-biotic Earth. The problem is that what that chemistry was is also a speculative theory. So you don't really know what to limit yourself to.

The modern version of the cell membrane is studded with protein structures that serve as signaling mechanisms for communication with other cells, as receptors to recognize the presence of molecules of interest (both beneficial and harmful) and to communicate the cell's own identity to others, and some that serve as portals which allow selective water-attractive molecules to pass through the membrane. It is clear that this protein machinery was not part of the first membrane that housed the first metabolism because these proteins require the full modern genetic machinery of DNA transcription and RNA translation to be synthesized. Also, the modern lipid bilayer is composed of phospholipids, a molecule with one head and two tails, and the growth of the cell membrane, before subdivision, is triggered by similar synthesis of more phospholipids inside the cell.

In Jack Shostak's lab across the river at Massachusetts General Hospital (I live in Cambridge), researchers have been investigating the properties of a much simpler fatty acid lipid with one tail that could have formed naturally on the primitive Earth. These also self-assemble into bilayers, micelles, and vesicles in water. Furthermore, the vesicle formation can be catalyzed by the clay *montmorillonite*, a kind of substance that might have been hanging around on primitive Earth. This is significant because this same clay has also been shown to catalyze strings of RNA from single nucleotides (the significance of this will emerge in the next section). The vesicle structures of the more primitive fatty acids are a little less stable than with phospholipids, but this could have been an advantage in the beginning. Individual molecules in the membrane will flip back and forth between the two leaves of the bilayer, making the structure more dynamic. This could have made the membrane more permeable to single RNA nucleotides, allowing them to get in to add to the original stock of evolvable material. Such nucleotides can't cross the border of our more modern phospholipid bilayers.

How to Reproduce a Metabolism

The simple budding off of lipid-bound metabolisms as they grow is a kind of reproduction, but not a kind of reproduction that would support the type of life we have come to know. Simple, wild type metabolisms that arise by accident would just be cloned, and thus repeated with more frequency. There would be more of them, but they would never improve on the original accidents. If the concentrations of the various catalyst molecules are not uniformly distributed throughout the cell volume at the time of subdivision, neither half will work like the original storm. Most likely, neither half will work at all. On the other hand, if the subdivision manages to produce two functionally identical storms, neither of them gets any better. The upside is capped; the downside is unlimited. For things to get better, we need two things. First, there needs to be some discrete, static memory of the key molecular catalysts encoded somewhere in the cell, in a form that does not get compromised by the storm, and this encoding needs to get duplicated with exacting precision, once, at subdivision time. This static catalytic backbone functions to normalize the daughter cells back to the original success, even if the storm is not evenly distributed at first. Second, this encoding of the metabolism needs to have enough slack built into it that small errors in duplication *mostly* denote the same catalysts, because *there will be errors.* That's how natural processes work. This tolerance for error allows past successes to be preserved as long as the copying is mostly correct. In the rare cases where these errors make a difference, they must make a very small difference, a slight tweak of the metabolism in either a positive or negative direction. It is the possibility of these small, unlikely, positive errors that makes evolution work. The occasional, accidentally better metabolisms will hang around longer. The occasional, accidentally worse ones will die out. The downside is capped; the upside is unlimited.

Unlike with metabolism, where the mechanics of metabolism change as we scale up from single cells to whole animals, the mechanics of reproduction remain unchanged from the single-celled form – cells divide (after first combining, in sexual creatures). It just wouldn't be feasible to do this with whole animals. If you cleave a human down the middle, you get two halves of a dead human. So when it's time for two humans to reproduce, each has to contribute a single, designated cell to do it for them, the old-fashioned way, then grow the fused cell back up to a human by repeated division. Those single, designated cells have to encode not only the entire catalytic backbone of each of the eventual 220 different cell types, but the entire developmental timing sequence for how the whole thing will unfold. We are all familiar with this concept nowadays in the form of DNA, those little twisted ladder things in cells that determine whether you get your grandfather's nose, or perhaps some genetic disease, or were at the scene of the crime. But it wasn't that long ago when this all just seemed like magic.

We have known about the general idea of genetics for hundreds, and even thousands of years. Plant and animal breeders have always been familiar with the notion that offspring inherit discrete features from their parents, and that the features of each of two parents are in some sense combined in the offspring. An abstract unit of inheritance was hypothesized to exist, and even named *gene*, well before anyone knew what its physical realization inside a cell might be. Until microscopy improved to the point that scientists could discern the fine structure inside cells, cells were thought to consist mostly of a continuous, oozy stuff called protoplasm, because the insides looked very fluid and homogeneous at any resolution then available. Somehow this ooze could reproduce itself and embody discrete traits that could be discretely combined. It was a great conceptual problem imagining how this continuous stuff could harbor such discrete properties, and even harder to imagine how the continuous ooze inside a fertilized ovum could contain all the information necessary to direct the unfolding development of a mature animal. It seemed as if something like a language would be required to hold all of this information, and a very powerful language at that. You can see a classic creationist argument forming here, and one was often advanced. Since it is impossible to record something as complex as the detailed fate of an organism in a liquid, the requisite information must exist somewhere else outside the cell. Some external force must guide the ovum's development into an adult organism. Life is a miracle. Miracles require Gods.

Writing in Molecules

When the molecular structure of DNA was finally decoded in the 1950s, scientists were expecting something discrete and perhaps language-like, but were a little surprised at just how digital and precisely linguistic it turned out to be. The erstwhile homogeneous protoplasm turned out to hold a number of little molecular translating machines, and the genetic code, specifying the entire fate of the whole organism, was literally written in molecules in every cell. The idea of writing in molecules is not so far fetched. The trick is to produce strings of molecular structures, like letters from an alphabet, with enough combinatorial variety to represent something of interest. We need molecular words and molecular sentences. Some molecules do naturally form crystalline structures, where long stretches of individual molecules are linked together in repeating structures of arbitrary size, but the repeating structure of crystals is too regular to represent much information. The crystal is essentially the same word or phrase repeated over and over again. To record the details of an organism's metabolism and developmental life history, we need lots of different sentences able to represent unlimited combinations of traits. This type of combinatorial variety can be found in molecular structures known as *polymers*. Polymers (*poly* = many; *mer* = part) are sequential chains of molecules (monomers), linked together by covalent bonds. When the monomers are all of the same type, as they are in plastics, we have the unhelpful monotony of crystals. When the monomers are drawn from a class of different molecules, though, we get the arbitrary expressiveness of a language. You can think of a polymer as a string of molecular symbols. When there is only one symbol in your alphabet, your symbol strings are limited to A, AA, AAA, and so on. If you allow strings up to length three, you can say at most three things. If you have two symbols, you can say A, B, AA, BB, AB, BA, AAA, AAB, ABA, ABB, BAA, AAB, BBA, BBB – fourteen things with strings up to length three. In general, if your alphabet has K different symbols, a string of N symbols can express K^n different meanings.

Life, it turns out, has three languages written in polymers. The DNA language (deoxyribonucleic acid) builds its strings out of four kinds monomer, each one composed of a chemical base connected to a piece of deoxyribose sugar that helps snap the sequence together. The four bases are *adenine*, *cytosine*, *guanine*, and *thymine* (abbreviated A, C, G, and T). The RNA language (ribonucleic acid) also has four symbols, but uses ribose sugar for the backbone, and *uracil* (U) instead of *thymine* (T) as one of the monomers. RNA is more like a different dialect of DNA than a distinct language. (The British say 'lift' with a British accent; Americans say 'elevator' with an American accent. DNA says 'thymine' with a deoxyribose accent; RNA says 'uracil' with a ribose accent). The third language, the language of proteins, has 20 types of monomer, all amino acids. The typical strings of the DNA language are quite long, about 240 million nucleotides (individual monomers), for instance, on the largest human chromosome, but these long strings are more like books than sentences. And the book is more like a reference work, such as a dictionary or encyclopedia, than a novel. It is a large conjunction of sentences, mostly unrelated to each other, that just happen to be written in sequence, rather than a narrative that flows from start to finish. The strings of the protein language, by contrast, range in length from about 50 to several thousand monomers. These are much more like sentences. Each one describes, and eventually becomes, a particular protein.

The Languages of Life

Now, suppose you set out to design some molecular languages like these to solve the heritable reproduction problem – some set of polymers whose strings of monomers could precisely capture the memory of a whole metabolism and reconstruct it later in a different cell. In addition to the languages, you would also need some fairly exact copying process for them, and some translation processes for them, and some manufacturing processes that could turn recorded names for metabolic parts into the parts themselves. This would be a non-trivial task even for an intelligent designer. But suppose there just happened to be some naturally occurring molecules hanging around the environment whose chemical properties give you all of this for free. Well lucky you, lucky us, lucky for life, there are!

First let's look at what metabolisms *are*. When you get up really close to one of these biostorms, you see lots of medium sized molecules randomly bouncing off each other but occasionally finding partners whose three-dimensional shape and electromagnetic charge fits and attracts some part of their own. They join together to make bigger shapes with different charges. Some of these aggregations will mix the charges in such a way to cause the whole pile to split apart along different boundaries, creating some new smaller shapes. And on and on it goes. The basic shapes are mostly molecules called *proteins*. What makes a particular metabolism unique is the particular variety of proteins involved (and perhaps their relative concentrations). Put a bunch of molecules with these same shapes and slots and charges together in a mix and they will soon sort themselves into doing the same collective dance. So to remember a metabolism, you just need to make a list of the proteins involved (mostly). If the dance requires twice as much of some particular protein, just list it twice.

So how do we form the names of the proteins in molecular words? Well, it turns out that proteins are *made* out of their own names! Proteins are composed of amino acids. Amino acids form polymer chains by linking to each other in strings. A string of these amino acids is literally a list of the contents of a protein, including the order in which they will be assembled. These amino acid monomers have strong attractive affinities that cause them to link side by side in strings. But they also have other attractive and repulsive affinities among their various kinds that cause them to fold up into three-dimension shapes once they first align in series. So why not just store a list of these pre-protein, string-of-acids names for the proteins as the canonical protein list to be remembered? Because their linear, string state is unstable. The string-of-acids wants to become a protein soon after it is formed. This is a good thing, because it does away with the need for a separate protein manufacturing process from the list of constituent acids. The list automatically becomes the protein. Just say it, and it shall be done! A similar problem prevents using the folded-up forms as names of themselves. Proteins want to react with other proteins. That's what they do. So they won't sit still in some preserved collection somewhere.

That's why we need a different language to record the names of the proteins. We need some kind of molecular symbols that easily form polymer strings, and then *stay that way*, resisting attempts to break up or fold up. It would also be nice if these strings somehow facilitated copying, and even better if they could provide some sort of error checking during this copying. Well, nature has a molecule that does all of this: DNA. It's four symbol molecules, A, C, G, and T, easily join together, side to side, in strings of indefinite length with the help of attached sugars and phosphate groups. But unlike strings of amino acids, the monomers don't further attract and repel their neighbors in the string. So the string *stays* a string. What's more, the four symbols of this molecular alphabet attract each other in pairs. A likes to bind to T, and C likes to bind to G. This attraction is not so strong that neighboring pairs will bend the string to get next to each other, but strong enough that free floating monomers will bind to the heads of their favorite partners in an existing string. So a single strand of DNA left in the presence of free monomers will quickly become a double strand as the complementary partners snap into place. Each side of this new double strand encodes essentially the same information with the base pairs reversed (this is the ladder that you so often see in DNA cartoons).

A message encoded in this default, double-strand form is even more stable because all of its symbolic base pairs are now bound to one of their favorite partners in the middle of the structure (the rungs of the twisted little ladder), with the sugar and phosphate backbones facing the outside (the sides of the ladder). All of the easy sites for reactions are now blocked. And because the message is encoded twice, once on each side, there is a backup copy available in case something goes wrong with either side. Lost information in one string can always be reconstructed using the other as a template. When it comes time to copy this double string, all you have to do is unzip it down the middle to expose the symbol molecules again. Free floating monomers will swoop in to repopulate their favorite pairs on each side, and voila! You have two exact versions of the same double-sided message. Gotta love that DNA.

The last trick is to map the unstable names for proteins in the amino-acid-string language onto matching names in the more stable DNA language. The protein language has 20 symbols (for 20 amino acids), but the DNA language has only 4 symbols. So we have to use more than one DNA symbol to form each amino acid name. Two DNA symbols per acid are also not enough because that would allow only 16 distinct names. So we have to go to three. That's enough for 64 names. Since we only need 20, that leaves 44 extra names on the DNA side, so there is extra expressive power left over. Nature has decided to use up 41 of the 44 extra denotation slots by making some of the 64 possible names redundant. ATG, for instance, denotes only the amino acid *methionine*, but GGA, GGC, GGG, and GGT all denote the same amino acid, *glycine*. This redundancy builds in some tolerance for error, because any copying mistake in the last letter will still denote *glycine*. Proteins themselves also embody some tolerance for error. Even if you get one or two of the amino acids wrong, many proteins will still fold up into shapes and charges that behave approximately like the intended one. They might catalyze a little less efficiently, but they still work. They might also just happen to catalyze a little more efficiently, creating the lucky accidents that evolution depends on.

With this translation scheme, a list of the amino acids needed to build a protein, and a list of all of these lists that make up a whole metabolism, can be represented and stored as a list of lists of triples in the DNA language. The DNA list (in double form) is stored just once, in one place, in the cell and provides an exact memory of that cell's particular metabolism. This is what gets copied, once, at reproduction time and passed on to the daughter cell. So you don't have to solve the harder problem of subdividing the metabolism with equal and representative concentrations of proteins in each half (perfectly cleaving a biostorm). The canonical list of catalysts is frozen in the DNA and will bring each side back into the proper balance.

The actual mechanics of translating the DNA back to proteins is both a transcription and a manufacturing task. For this, nature uses the third language, RNA. It has almost the same naming scheme as DNA (with T changed to U), but its chemical properties are somewhere in between DNA and proteins. It is fairly stable in strings, but not as stable as DNA. It runs around the cell mostly in single-strand form, so it is more suited to holding temporary messages. Its strings will occasionally fold up in hairpin turns and other shapes that give it some of the reactive, catalytic capabilities of proteins. Its fragments shuttle back and forth between the DNA library and the metabolic storm, relaying the ingredients of proteins, folding into its own little widgets which interact with proteins, and sometimes combining with proteins to form the really big, stable molecular machines, such as the ribosome which converts RNA messages to protein messages.

Here is where we cash in the significance, noted earlier, of a clay catalyzing both primitive lipid vesicles and RNA strands. Because RNA is both a self-reproducing, fairly stable language for preserving information (like DNA), *and* able to fold up and become a three-dimensional catalyst (like proteins), it is hypothesized that life first got started in a simpler "RNA world" in which RNA supplied the original molecular basis for both reproduction and metabolism. Over time, DNA took over the bulk of the information role because it was better at static things, and proteins took over the catalytic role because they were better at dynamic things. RNA is now the middleman, but its jack-of-all-trades fingerprints are still found in almost every aspect of the cell's machinery.

3 | The Origin of Life

Darwinism asks us to believe that one day there was nothing but mud and ooze, and the next day there was life ...
    – Ben Stein, *Expelled.*

The origin of life, the transition from non-living matter to living creatures, has always been viewed as one of those quintessentially inexplicable events that borders on the miraculous. If you are religious, or otherwise believe that possession of a soul is what makes a physical body alive, the miraculous origin of life, in the form of ensoulment, must happen in at least three different contexts: the origin of the first life from non-life, the origin of the first human life, and the origin of each subsequent human life. If you are simply a vitalist, and suspect that some special extra ingredient gets added to inanimate matter when life happens, you only have to account for one event: the original non-life to life transition. Since life begets life as part of its normal operation, the special vital force is simply subdivided, or shared, or otherwise passed on to each subsequent life. The only really anomalous begetting is the first one. How did the vital force get into matter in the first place?

Souls, however, complicate the story. Because they also encompass the notions of moral responsibility and individual identity, there is a need to distinguish certain grades of souls, and to assign a distinct, new soul to each new human. So life begetting life is also an event that requires special intervention. Since bacterial souls (or whatever it is that they have that makes them alive) are of a fundamentally different kind than human souls, even if you believe in evolution, you are committed to some discontinuous singularity at the point that the first human one arises. You also have to account for the creation and installation of a new soul each time another human is born. Somewhere between the initial gleam in the eye that starts two unrelated haploid cells on a collision course and the birth of their fully gestated union, a new soul needs to be associated with the integrated collection of trillions of diploid cells that results.

The soul theory, of course, grew up without the benefit of our current understanding of molecular biology, which is why it is so hard to retrofit onto what we know now. If you are not burdened by the need to insert souls into biology, life forms a very natural continuum from humans back to bacteria, with the origin of each new life being an approximately faithful reproduction of a previous life. Somewhere, though, the great chain of being must end at a point where the first life came about without benefit of a previous one. Abiogenesis, the spontaneous origin of life from non-life, has always struck even men of science as an extraordinary proposition. Until the middle of the 19th century, it was thought, even by some scientists, that the spontaneous generation of life from non-life was a contemporary, observable phenomenon. Maggots and molds were thought to be spontaneously generated from dead vegetable and animal matter. We now know, of course, that our predecessors were merely witnessing normal life-begetting-life events. They just couldn't see the begetters. Once these phenomena were pushed into that more familiar category, they were no longer extraordinary. But during the centuries in which abiogenesis was thought to be directly observable, it still boggled the mind. We are simply unprepared by our folk chemistry for the crossover from mineral to either animal or vegetable. It's somehow not natural. Something must intervene to provide the special essence of life. What that special something is has changed over the years to accommodate our changing criterion of life. For a long time, we identified life with the respiration of air, because every living creature we were aware of breathed. When an animal stopped breathing, it died. If a newborn didn't start breathing – soon – it failed to live. So we came up with the notion of the breath of life. After fashioning their physical bodies, creator Gods turned inanimate things into living things by breathing the breath of life into them. (Fish were somehow not accounted for). In Mary Shelley's *Frankenstein*, electricity was the special ingredient that animated a surgically composed collection of body parts. In general, autonomous motion – as in animation – seems to be the essential feature that separates the living from the non-living (the quick from the dead), and thus the crossover from inert matter to animated matter is the extraordinary event that needs explaining.

The Animation of Mud

Our modern definition of life has the benefit of many more observable life forms to generalize over. Following the life-begets-life lineage backwards from large, visible things that are clear examples of life, we can trace all the way down to bacteria. Along the way, we have discarded such things as arms, and legs, and oxygen respiration as inessential features, and ended up with the big three: metabolism, containment, and reproduction. The curious thing about this definition of life is that there are several kinds of things that possess only some of these traits, and thus don't qualify scientifically as life, that might have qualified under the old "autonomous motion" criterion. Viruses, for instance, have an encasing membrane, a set of genes, and a very successful and prolific reproductive strategy. But they lack a metabolism. They must temporarily hijack a host cell's metabolism, and some of its reproductive enzymes, to get themselves reproduced. You might even say that they have a purpose in life that they are quite good at – rampant reproduction and evolutionary survival – except that we don't regard them as having a life to begin with.

All of this testifies to just how arbitrary the exact criterion for life is. We identify life with a cluster of characteristics that seems to capture the greatest number of instances that we already regard as life, and we don't worry too much about the boundary cases. That is because the exact event that first tipped non-life over the threshold into life is of no great significance to science. If life is defined as the confluence of metabolism, reproduction, and membrane containment, then the first life arose when these three features, which each have their own independent reasons for forming, first got together – in any order. It doesn't really matter which one came first, or whether they each arose in parallel, or whether certain combinations mutually developed first. What matters is that each of these quasi-biological structures was a likely natural consequence of the more primitive chemical and physical structures that preceded it. Since we have already accounted for how to make each of these structures out of a few molecules with just the right chemical properties – the spherical, lipid bilayer, auto-catalytic reaction cycles of proteins, and self-reproducing polymer syntax – we can observe the erstwhile momentous event of the origin of life in this chapter simply by saying "and then they got together."

If you were waiting for the magic moment, or the special ingredient that turns non-living matter into living matter, you will be disappointed. All of the parts are already here. The actual flipping of the switch from non-life to life was a huge anticlimax in the grander scheme of things. In the scientific story, the magic moment never arrives. It was just an ordinary day on the primitive Earth, with the usual bunch of polymers banging into and catalyzing each other. Almost without notice, some group of them achieved, at least temporarily, the requisite three-way combination. No one applauded, no choir sang, no celebration ensued. There probably wasn't even a single event. More likely, there were many such events – a lot of near misses, a lot of temporary starts that didn't last, a lot of advancings and retreatings – before life took hold permanently. But the world looked pretty much the same immediately before and after. And this is how it should be if we are truly explaining things. Successful explanation, as we will often say in this book, proceeds in small increments. Miracles appear to be required only when you try to take too large a step all at once. This is not to say that life is not an impressive phenomenon. But as is typical with impressive phenomena, it is impressive because it emerges gradually somewhere on a long continuum. What is impressive is the contrast between two widely spaced points on this continuum, between something that is not life-like at all and something it eventually transitions to that is the very epitome of life. So why were we expecting more?

As with most of the mysterious gaps that we think we see in nature, this one is due to our inability to see the continuum. We only see the end points. If you take away our special knowledge of microbiology, which is not observable to most people anyway, our perception of the transition from inert matter to living things is about the same as it was for our ancestors who came up with the breath of life theory. When you gaze out on a quiet, peaceful meadow, next to a still pond, under a motionless blue sky, you wonder how the noisy, busy cacophony of life could have arisen from such a silent, motionless beginning. When you try to imagine all of that inert matter in front of you becoming the very lively bodies of familiar animals, you are at a loss for a transition story that doesn't inject some external source of liveliness into the inertness. But you are wrong about what you are witnessing, on two counts.

Life All the Way Down

First, you missed the busy cacophony of life that was already in full swing. Right in front of you, epic battles for survival are raging between billions and trillions of combatants, many using sophisticated chemical warfare to subdue or contain their opponents. There is also plenty of harmonious life activity transpiring. Agents are busy trading commodities, collaborating in cooperative networks, and working together in elaborate societies to accomplish joint benefit. They just happen to be too small to see. Each gram of soil in that meadow contains about a million fungi, and from 100 million to three billion bacteria. These are the primary combatants. Bacteria and fungi compete with each other for the resources to grow and reproduce, and keep each other in check by lobbing an endless variety of small molecule poisons at each other. Since these battles have been raging for billions of years, the chemical arsenals have had plenty of time to evolve in relation to their targets. For every species of microorganism, there is probably already a specific antibiotic manufactured naturally by one of its competitors. The bacteria and the fungi also inhabit the air under that clear blue sky, in the form of spores hitching a ride on the breezes in search of new homes. About a million bacteria make a home in every milliliter of that still pond. Many bacteria also make a home on and in you.

The relationships between these extremely small creatures that you don't see are often more complicated than the simple peer-to-peer model we are familiar with at the large level. Many make their living as parasites, organisms whose natural environments consist of regions of much larger organisms, such as yourself, that are not aware that they are providing hospitality. Many of these parasites have lifecycles that span multiple intermediate forms that have evolved to track a dependable journey through more than one kind of animal host. For example, the single-celled parasite *Plasmodium*, which causes malaria in vertebrate hosts, lives in the salivary glands of a female mosquito in a sickle-shaped form called a *sporozoite*. When the mosquito bites a vertebrate, such as a bird or a human, the sporozoites ride the mosquito's saliva into the hole in the victim's blood vessel. From there, they migrate through the blood stream of their new host to its liver, where they enter liver cells and divide into thousands of new oval-shaped forms called *merozoites*. These forms reenter the blood stream in order to invade red blood cells, reproduce more of their kind, and then burst out of the blood cells to infect more blood cells. Eventually some of the merozoites will reproduce a new sexual form called *macrogamonts*, which wait around in the blood for another mosquito to bite. When the bite comes, they are sucked into a new mosquito's gut, where they will mate to produce a new round-shaped offspring called an *ookinette*. This form will eventually divide into thousands of the sickle-shaped sporozoites that will, in turn, migrate to the mosquito's salivary glands.

Needless to say, this arrangement is not a particularly happy one for the vertebrate host. But just as often, the relationship between the little critters and the bigger ones is symbiotic. In return for living inside a larger host, the smaller organisms provide some vital service that the host needs to survive. Metabolism is probably the most common form of this symbiosis. The really tiny organisms have a much greater variety of methods for breaking down large molecules into smaller ones. Most large organisms harbor colonies of smaller ones in their digestive tracts to break down the things that their own protein enzymes are not capable of handling. Many termites, for instance, which are famous for being able to eat wood, cannot actually digest it. The Northern Australian termite *Mastotermes darwiniensis*, for example, chews the wood into very small pieces, but it lacks the native biochemical ability to digest the cellulose and lignin that make wood, and other plants, stiff. There is a lot of organic food content in these compounds that is unavailable to most animals because they lack the cellulase enzymes necessary to break it down. Various microorganisms do possess these enzymes and produce acetic acid as a waste product, which is nutritionally useful for the larger animals. Approximately one third of the body mass of Mastotermes darwiniensis consists of millions of microbes living in its hindgut that perform this beneficial service. The termites could not survive without the embedded microbes.

But like the Russian matryoshka dolls that are nested inside each other in smaller and smaller forms, some of these microbes are themselves symbiotically dependent on their own, much smaller, embedded microbes. One of the largest, and most interesting, of these little symbionts in the termite's hindgut is a protozoan called *Mixotricha paradoxa*, which itself contains hundreds of thousands of much smaller bacterial symbionts to which it subcontracts the digestive task. Mixotricha is famous for being mistaken on its original discovery in the 1930s for a single organism with a strange (paradoxa) combination of cilia and flagella hairs. Its original discoverer lacked the modern technology of electron microscopy and could thus not view it in motion or at the finer resolution available today. The "hairs" turned out to be about 250,000 independent spirochete bacteria attached to little slots in the Mixotricha body via another species of pill-shaped bacteria. These once free-living bacteria had evolved to become the symbiotic workers for the larger Mixotricha cause, rowing it around the termite gut so that other bacteria contained inside could do their wood-digesting thing. In all, Mixotricha represents a combination of smaller critters from five previously independent genomes that have become permanently locked into a cooperative endeavor, which is itself just a cog in the larger symbiotic endeavor of the back end of a termite.

So one reason that we find the emergence of life surprising is that we don't really see much of it. What we take for inert matter, at our size, is often itself composed of lots of life – competing life, cooperating life, life living entirely off of other life, life inside life inside life, all the way down. We find the gap between inert matter and living creatures so great because the representative samples of each that are observable at our size are so dissimilar. The meadow isn't much like a rabbit. The pond isn't much like a fish. The air isn't much like a bird. If we were able to partition the seemingly inert stuff into the really inert stuff versus the much greater quantity of living stuff that just looks like inert stuff, the contrast would not be as great. Our viewing life on Earth from the meadow is a lot like aliens viewing human life on Earth from a near orbit in space during World War I. As trench and chemical warfare raged across Europe, as societies and nations hummed along in trade and commerce, in exploitive and cooperative endeavors, Xgsrt says to Brogf, "My it looks so peaceful and still down there. What a pretty blue world. Too bad it doesn't contain any apparent life." We are like Horton the elephant, too large to hear the Whos. The vast majority of the 3.5 billion years that life has existed on Earth was spent evolving these little guys. The larger plants and animals that we can see are a relatively late elaboration of these more basic themes of life. The purposeful motion that we are looking for was already in full swing a long time ago. It just got bigger.

Motion All the Way Down

But even if you transfer your wonder about the animate/inanimate gap to the micro scale, trying to imagine how purposeful *small* motion could have sprung from motionless *small* matter, you would be wrong on the second count that we left hanging from above. There is still no abrupt transition to wonder about. As we have seen in the previous chapter, there is motion all the way down. There is effectively no such thing as motionless matter. Stillness is a perceptual illusion of large beings. Even below the atomic level, which we take to be the ground floor of matter here on Earth, the fundamental forces keep the subatomic particles in a frenzy of quantum jitters. This peripatetic motion continues up through the molecules, which shimmy within due to Brownian motion and react with their neighbors due to electromagnetism. What changes as we go higher in organization is the *coherence* of that motion. It begins to form patterns, then cyclical patterns, then cycles within cycles that implement goal-directed survival behavior that allow cells to actively prolong their own existence. The transition to the motion of life is a gradual corralling of the ever-present motion of matter into the circular kinds that lead to repeatable persistence.

If the concept of vivified mud still seems like an odd way for life to begin, try this perspective. We live in a universe in which the matter is in constant motion. We, and the other large things we are able to perceive, are composed of so many embedded layers of this motion that from a distance our collective motion looks like stillness. We see only the coarse-grained, summed-up motions of arms and legs and such, and take this to be representative of coherent motion in general. The motion of life is actually motoring along at various levels in all three of our macro categories: animal, vegetable, and mineral. The motion of the animals percolates all the way up to the top level at speeds we can observe. The motion of the vegetables at top level is too slow to observe (without time-lapse photography), but their constituent small motions are much faster. Mineral motion? That black stuff in your shower – that's life. That black stuff you call dirt – that's life. The blue in blue cheese, the bitter in beer, the tang in yogurt, the sour in pickles – life, life, life, life.

When you restate the problem this way, as one of transitioning not from stillness to motion, but from motion to coherent motion, then from small coherent motion to large coherent motion, there is no expectation of a special point event. Coherence is a holistic notion, something that arises gradually, in degrees, to a large collection of things. Life doesn't happen suddenly. It fades in.

4 | That Evolution Thing

I didn't believe in the theory that human beings – thinking, loving beings – originated from fish that sprouted legs and crawled out of the sea. Or that human beings began as single-celled organisms that developed into monkeys who eventually swung down from the trees.

– Sarah Palin, *Going Rogue*

To the rest of the world, the highly publicized 2005 Federal trial concerning the teaching of evolution in the public schools of Dover, Pennsylvania (Kitzmiller v. Dover Area School District) seemed oddly anachronistic. Hadn't this issue already been settled almost a century ago in the equally famous "Scopes Monkey Trial" in Tennessee? It was hard to fathom how a country so renowned for leadership in science and technology abroad could be regressing so badly at home. This is the view from 35,000 feet, though. It's missing some details. Despite the impression left by the 1960 film version of *Inherit the Wind*, the advocates of religion won the earlier decision, and the teaching of evolution remained illegal in the State of Tennessee for another 42 years. In the Dover trial, the advocates of science won the decision, protecting the long since legal teaching of evolution from attempted abridgment by religion. The Law has come around. It is the views of some ordinary citizens that have lagged.

High school science education trails the actual practice of science in research universities by some distance. There wasn't much evolutionary content in the textbooks to teach until the 1960s, and even then, it was routinely ignored by teachers in the Midwest and South. As the content did emerge, it set off waves of new statutes by school boards and state legislatures, primarily in the South and Midwest, that led up to the Dover decision. There was a characteristic lifecycle to each wave. The statute would begin at a local level as an attempt to impugn evolution by mandating the equal teaching of Biblical creation. As the statute moved higher in visibility, it would be rewritten to obscure its religious motivation, and its supporters would be coached on what not to say in public hearings if they wished the statute to pass constitutional muster. This was a difficult strategy to orchestrate because the grass roots sponsors were often openly proud of their religious motivation and did not see why religion should not be taught in public schools. Each wave would ultimately crest in a defeat at the US Supreme Court as the Justices routinely sniffed out the religious content and established yet another criterion for what counts as science and what counts as religion. This affected a kind of evolution-by-judicial-selection in the statutes themselves. Proponents gradually learned that you can't really say what you want to teach – religion; you must limit yourself to saying what you don't want to teach – some aspect of science. And your reasons can't be that the science runs counter to your religion, they must be a stand-alone criticism of the science itself. Not a problem. Ordinary citizens, local school boards, and state legislators who have never set foot in a real science lab in their lives feel they know all about what is wrong with evolution. It is an unfortunate, aberrant stepchild of respectable science, a pernicious philosophy of atheism – Darwinism – that leads some scientists to see things that just aren't there. These guardians of public education aren't against science in general, just the few bad parts.

Among the common things that these citizen scientists know (somehow) is that "evolution is just a theory." 'Just' is the operative word here. It serves to set evolution apart from mainstream science which is presumably *more* than just a theory. You could drop evolution, and the rest of science would get along just fine. So there would be no material disruption to the rest of high school science education if you excised just this little part. Unlike other parts of science, they know, there is very little evidence to support the theory. It is mostly fossil evidence, and that is mostly of dinosaurs. The rest is pretty sketchy and open to a wide range of interpretation. Some see humans, some see apes. Who's to say? There are also lots of gaps in the fossil record, so they've heard. Pretty slim pickings on which to base a theory as far-reaching as evolution. In fact, this is what sets evolution apart from the rest of science. It is uniquely historical. It is a guess about what happened in the deep past. It cannot be observed in the present. It can't make any predictions about what will happen in the future. You can't do any experiments to confirm or refute it. It is all the result of the imagination of one man, Charles Darwin, who has gained an outsized influence on the thinking of modern scientists. This was once true about Sigmund Freud, but scientists eventually got over that. Someday, it is hoped, they will get over this Darwinism too.

It is ironic that many of these sentiments are listed as "findings of fact" in the proposed statutes, though no one appears to have consulted real scientists. Even more curiously, they are almost uniformly non-findings of facts – findings of gaps, omissions, and lack of evidence, after diligently not looking for any. Academics know how hard it is to substantiate a claim that "there is no known evidence for X." They usually hedge by saying instead that "I am aware of no evidence for X in the literature," implicitly offering their reputation as an expert in the field, someone who diligently reads the relevant literature and thus would have some reasonable expectation of knowing what was out there. "There is no evidence for X because I didn't find any, and I didn't find any because I didn't look for any" is not a claim worth making in any forum in which people genuinely care about finding the facts. But it works when the majority of your constituents are already disposed to believe the "facts" you've "found." There are an estimated 97 million Americans who don't believe in evolution, so you are bound to get a sympathetic hearing.

The Other Evolution

If these finders of fact were to spend some time in a modern research university, they would discover that their concept of evolution is a little out of date. They would be hard-pressed to find a copy of Darwin's *On the Origin of Species* in a science class or lab. They would not encounter the term 'Darwinism.' Darwin is just one of the many members of the scientific hall of fame, up there with Copernicus, Galileo, Newton, Einstein, and many others who first contributed seminal ideas to our evolving, integrated theory of the natural world. They are revered not because they founded schools of philosophy about which present scientists debate, but because they contributed initial theories that got other, subsequent theories going down a productive path. Much of what each originally proposed turned out to be a little bit wrong in the details, as one would expect. But each proposed enough that was right that their original theories could be adapted to what was discovered later. The lasting scientific value lies in the theory, not the man.

Newton's case was the rare one from early science where the original theory was advanced with a rigorous mathematics right from the start. You knew exactly what Newton was proposing and how one would go about measuring and testing it. More common is the development of the theories of thermodynamics that started out as engineering ideas concerning the efficiency of steam engines, and then gradually acquired a mathematics that enabled them to be precisely measured and tested as laws governing the transfer of energy at all levels of organization from molecules to galaxies. Evolution was like this. Darwin had an idea, not yet a mathematics. The ideas were formed around the large, multicelled plants and animals, and their few fossils that were visible at the time. He didn't have a mathematics or a theory of the fine structure of genetics and cells that underlie visible mutation and adaptation. But he had the basic outline right. His outline has subsequently been filled in with the precise molecular underpinnings and a mathematics that describes the precise dynamics of evolutionary processes. Like the formal principles of thermodynamics, these mathematical models apply to processes at all levels of organization, from bacteria to immune systems to ecologies to social behavior to language. A process has an evolutionary dynamics if it has a collection of types of individuals that replicate at some rate, mutate at some rate, and compete for survival at some rate. From initial conditions, you can sometimes prove that certain stable equilibria will eventually arise among the resulting populations, or that the process will oscillate or cycle through a pattern of temporary equilibria. For models too complex for analytical results, you can often approximate future states with computer simulations, like one does with weather. Evolution is now one of the main backbone theories that underlies and integrates whole branches of science. The dynamics of large metazoan speciation, Darwin's original application, is just one of its many applications.

The action in modern biology, particularly where evolution is concerned, is now at the molecular and cellular levels. These are the atoms of life. The larger life that we see is just an elaboration and composition of the principles of life at this smaller scale. Biologists have relocated down here for the same reason that physicists moved from billiard balls down to elementary particles. Paleontology still goes on, but because only the hard skeletons and shells of larger organisms fossilize, it doesn't tell us a lot about these smaller units. But now we have something much better. High speed, whole-genome sequencing allows us to look in exquisite detail at the living fossils in extant DNA and RNA and protein complexes. The evolutionary history of all organisms, who begat whom and approximately how long ago, can be computed by comparing relative changes in the ribosomal RNA across species. Traditional taxonomies based on observed morphology have given way to *cladistics*, where the computer places species into a common group (a *clade*) if they share some unique traits that first emerged in their last common ancestor. This allows likely evolutionary histories to be computed (trees of inheritance, or *cladograms*) from hypothetical ancestors for whom we don't yet have fossils.

This has radically altered our understanding of the organization of life. We now know, for instance, that the vast majority of all the genetic diversity on Earth is among the single-celled creatures. We now know that there is significantly less genetic difference between a human and a potato than there is between the bacterium that causes tuberculosis and the one that causes cholera. Whales are more closely related to hippos than hippos are to pigs (or any other ungulate). All *bilaterians* (multicelled creatures with symmetrical right and left sides) share the same basic genes for body plans from humans right down to sea urchins. We share about 98% of our genes with chimpanzees, 92% with mice, 70% with zebrafish, 44% with fruit flies, 26% with yeast. If the investigating committee for the school board were to walk across campus to the medical school, they would no doubt be surprised to find that the researchers studying causes and cures for cancer and diabetes and ataxia in humans are studying mice and fruit flies and worms and yeast. Same genes, same regulatory pathways, same molecular mechanics – but simpler, because these model organisms are more basic than humans and infinitely easier to experiment on.

Dealing with life at this smaller scale has also made it possible to observe evolution in ways that Darwin could not. At the large metazoan level, evolutionary changes in large features take too much time to observe in the laboratory. But evolution is still happening. It continues unabated in the never-ending war between the rapidly evolving little pathogens that cause our diseases and our own adaptive immune systems. Over longer periods of time, but periods still measurable in modern historical units, we can trace evolutionary changes in our human genetics for immunity to bubonic plague, sickle-cell anemia, and Lassa fever, and for lactose tolerance in adults. But when you pit smaller organisms directly against smaller organisms, whose lifecycles per generation are much shorter, like fruit flies, you can observe evolutionary change in the laboratory. When you deal down at the bacterial level, where reproduction can occur every twenty minutes or so, you can observe evolution in hundreds of thousands of generations in real time. You can experiment, measure, observe, predict, confirm and refute. It goes on every day in labs all over the world. But scientists aren't performing these experiments to confirm or refute whether evolution happens at all, any more than chemists experiment to confirm or refute whether thermodynamics happens at all. They want to discover more of the details about how it happens.

Fish Sprouting Legs

We opened the previous chapter with Ben Stein's skepticism, often parroted, about "Darwinism." Actually, evolution has nothing to say about the origin of life because it only deals with things that already self-replicate, but we know what he means. He is expressing skepticism about abiogenesis, both the very idea that mud could become life (unless a deity breathes life into it, of course, or perhaps fashions it out of a rib – now that's believable) and that it could happen so quickly (again, speed no problem for a deity). Sarah Palin's skepticism, that opens this chapter, covers the proper range of evolution, but offers a version of the same two incredulities: how could one living thing possibly transform into a completely different living thing, and how could it happen so fast? One difference, in the proper evolution case, is that even deities apparently don't do species-to-species transformations (surely they could if they wanted to). They create each kind separately, but they can still do it awfully fast. Granted, neither of these unbiased aficionados of science has any real interest in getting to the bottom of the theories they disparage, but they do express a common skepticism that even open-minded people have.

You, dear reader, are one of those open-minded people of course, so you have no agenda that prevents you from getting to the bottom of this. Still, with your ordinary, day-to-day human-being hat on, this whole sweep from single cells to humans is quite an extraordinary proposition. At the macro scale of human experience, the difference between mud and a fish, say, is quite dramatic. One just lies there; the other swims about purposefully. But at the micro scale, the transition from non-life to life was much less dramatic, as we've seen. The mud already contained the precursors of life. They just had to get together in the right combination. From the macro point of view, it would look a lot like mud turning into mud. That's plausible. Fish came much later. And it took about a half billion years for the biochemistry of the pre-life mud to get to the point where it could take this very small step across the line. Add another three billion years or so to get to fish. Fish took another quarter billion years to sprout their legs, and yet another quarter billion years to get to the loving humans. All of the constituent transitions occurred in tiny little biochemical increments at the molecular scale. They collectively added up to a great deal.

So the first key to understanding the full sweep of evolution is to forget about this "sudden transition" business. It's a red herring. 3.5 billion years is not sudden. Nothing happened quickly. No evolutionary transition from one living kind to the next took place on any scale that a human would regard as observable. Individual fish didn't sprout legs; individual monkeys didn't turn into humans. The very idea! Well that's a relief. You wouldn't want a scientific theory to depend on natural processes being able to do preposterous things like that, when even Gods don't do that.

But wait. Nature does do that. Remember caterpillars that turn into butterflies, and tadpoles that turn into frogs? The same organism. In real time. Hmm ... worms sprouting wings and fish sprouting legs. Right in front of us. Isn't it odd that nature couldn't possibly do in 250,000,000 years what it can manifestly do in about 58 days? We call this visible, real time transformation from one kind of animal to another *metamorphosis*. It is quite common in non-mammals, where embryonic development occurs outside of the parent. In mammals, the development progresses on the inside, so the transitional forms from single cell to whole animal don't have to be viable life forms outside of the womb. This is one of the reasons that we tend to ignore it, being mammals ourselves. We don't think of these intermediate things as animal types in their own right. But in the rest of the world, the stages have to survive on the outside as real animals.

The eventual frog begins as a single-celled egg, *outside* of the parent, that grows and divides into a fish-like form with gills and tail. It uses the gills to respire oxygen from the water and the tail to swim. Then skin gradually grows over the gills and they recede and disappear as lungs begin to form. About the same time "leg sprouting" starts, followed by arm sprouting. The critter converts from gill respiration in water to air respiration at the surface. Then it moves onto land and its tail atrophies. Now it's a full-fledged tetrapod. All vertebrate embryos go through a similar transition, even the ones inside mammals, like humans. They all reach a phylotypic stage, called the *pharyngula*, where they have gill slits and a tail. Then they branch off on different development paths to match their eventual end kind. Fish further refine the gills and tails and sprout lateral buds that become fins. In land vertebrates, the gills atrophy, lungs develop and the lateral buds become arms and legs. For creatures with final tails, the tail keeps developing. For creatures without final tails, the tail atrophies. We are vertebrates. This is how we develop. So if you find yourself doubting whether loving humans could possibly have evolved from single cells, even over 3.5 billion years, think about how it happens all around you inside your fellow humans, in about 9 months. All of those human traits that seem so implausible to have emerged from aggregations of cells during evolution are emerging yet again, over and over, as we speak (or read in this case).

There are lots of things about evolution that are hard to fathom, because there is no analog for them in normal human experience, but this is not one of them. It's not like astrophysics or quantum mechanics where you are asked to imagine something that you couldn't possibly experience in your lifetime. You don't have to use your imagination at all. The stunning transformations from one animal to another are happening right here, right now. Even if you are religious, there is no practical reason for your religion to conflict with evolution on this point. Since you know that the erstwhile offending transformations are taking place right now in the form of embryonic development, they must be possible. You wouldn't want to say that a God is not capable of pulling off this trick in the present. So if a God is orchestrating this in the present, behind the scenes, putting in souls, and consciousness, and love at just the right junctures, what's the point of denying that this same kind of thing was happening in the past? Gods could have done it then too, in the same way they are doing it now. Why would they have to switch methods in the modern world? If you don't have to deny evolution, then you don't have to come up with stories about why Gods planted all of this burgeoning historical evidence in fossils and genomes just to mislead modern scientists.

Biological Plagiarism

In many ways, embryonic development recapitulates, in compressed time scales, the mind-boggling evolutionary transition from single cells to large animals, giving us a living laboratory for how something like this could have come about in the first place. This transition is clearly a much harder problem to solve than the original origin of life. There is irony here. If reproduction of single cells had somehow required a recapitulation of abiogenesis – something like cells emitting separate versions of membrane, metabolism and genes, temporarily dipping below the level of life before putting them back together again to get the next organism – we would have had a working model of the origin of life in the present to study. But it didn't work out that way, and nature erased the transitional steps, so we just have to keep guessing how it worked until we manage to synthesize something like this in the lab. On the other hand, if nature had somehow figured out how to reproduce whole multicellular animals directly by binary fission, and erased the transitional steps, we may never have figured out the historical transition from single cells to humans. But because nature had to solve the large reproduction problem by reducing back down to single cells and emerging back up to animals, leaving lots of transitional forms along the way, we get to see how to do this in the present. We would probably never have figured it out by guessing in the lab. Much of what we have learned about evolution recently comes from this emerging field of evolutionary developmental biology, known colloquially as *evo-devo*, where the real-time development of embryos can be studied across species, cell by cell, gene by gene, and compared to the more traditional fossil record.

As you might expect, when you study development this way, you discover many traits that we share with mice and frogs and fruit flies and worms, such as our immune system, our nervous system, and our metabolic pathways, all of which speak to a shared ancestry. But you also discover the *useless* things we share with other creatures, the harmless mistakes, the inelegant designs, and the features we share with other species that are essential for them but useless for us. We harbor broken, or incomplete, or discontinued versions of their very functional genes and features. This is a very important concept to grasp if you want to understand evolution. Before we got a good look behind the scenes at how this stuff actually works, we used to marvel at the beauty and elegance of nature's designs. Organisms seemed to be perfectly designed to match their habitats. This perspective was once used as an argument for an intelligent designer. The accidents of nature could not possibly provide the foresight and planning required to fashion such perfect fits. But now that we see the scruffy, almost comical, way in which the parts are put together to achieve the end result, it makes much more sense that this stuff was fashioned by a bumbling, incremental, trial and error process with no foresight and no sense of elegance. It is full of dead ends, and backtrackings, and restarts, and repurposing of old parts until something works, with no imperative to clean up the messy process that got you there.

Perhaps the most ubiquitous of these is the pharyngula stage of vertebrate development. It makes perfect sense to make gills and a tail if your ultimate product is a fish. But why do this if the ultimate product is a tailless, land-based tetrapod, like humans? Why not just leave that part out, instead of first making it, then undoing it? This method would be senseless if you were starting from scratch to make a tetrapod, but if you were starting from a reasonably successful fish design, and started tinkering with the latter half of the process and managed to repurpose it into a tetrapod with lungs, well, that would make sense. The hybrid design might even come in handy to build a final fish that turns into a tetrapod, a la the tadpole/frog.

We humans lose almost all vestiges of our pharyngeal gill slits and tails by the time we pop out. We retain a "tail bone," the *coccyx*, at the end of our spine that forms from a fusion of the last four vestigial vertebrae as the erstwhile tail is reabsorbed into the fetus. The gill-slits portion of the embryo that was once destined for fishhood gets drastically repurposed in us so that the gill arch bones of the fish become our middle ear bones. But some of our final anatomy is still pretty fishy from an elegant design perspective. The aorta, the main artery carrying oxygen-depleted blood out of our hearts to pass through the lungs for re-oxygenation, first proceeds up toward the head before it abruptly hooks around and heads back down. What's that for? Well, back in the pharyngula stage, the developing aorta already courses up through the developing gill slits, because this is where blood gets oxygenated in the final fish. As our gill slits atrophy and lungs develop further down, in line with the modified breathing design, the source of final oxygenation gets relocated. So the developing aorta has to take a sudden U-turn and head back down, retracting its connection to the gills that will no longer be there. Why would you design something like this from scratch? You wouldn't.

Dinosaurs, you may have heard, didn't completely die out when the Earth was struck by an asteroid near the Yucatan Peninsula about 65 million years ago. It was total bad news for the land and water based species, but the few who could fly muddled through to become the ancestors of modern birds. Scientists had long suspected such a lineage because of a similarity in anatomy of the hind legs that hinge in the opposite direction from those of most tetrapods. We don't have any dinosaur DNA to confirm this, because DNA doesn't last that long, but there is an intriguing clue in the surviving embryonic development. Chickens apparently retain the genes for ancestral teeth. These never make it to the final chicken because in normal development these genes are suppressed. But a mutation, first discovered accidentally in 2006, can turn these genes back on, leading to embryonic development of bird teeth. The mutation is fatal, so the embryo does not last past about 18 days, but in that period it begins to grow teeth. Modern birds have no use for teeth, so there would be no point including this vestigial trait in their de novo development just to suppress it. But if it was already there in a repurposed design, well.

Bird teeth are a case where a vestigial gene was discovered when its effects were reactivated. More common are broken genes that are passed from ancestral species to descendent species without reactivation. Because the original is broken, i.e., it sustained a negative mutation in the ancestor that rendered a useful trait non-functional, but not so useful that its loss threatened the survival of the species, the mistake simply rides the DNA from one generation to the next. These can be discovered by comparing the genomes of different living species to find almost identical misspellings of DNA sequences that are functioning genes in other species when the letters are in proper order. Examples of these vestigial pseudo-genes in us are one whose correct spelling codes for a protein to synthesize vitamin C, and about 70 whose correct spellings code for odor receptors in the nose. Currently living primate species, such as chimps and macaques, which share recent common ancestors with us, have these same misspellings, and the similarity of the misspellings increases as the hypothesized ancestors get closer to us. Primates get by without synthesizing their own vitamin C because we can get it by eating fruit, and primates' improved vision has supplanted the need to distinguish as many different olfactory odors as our mammalian predecessors. But these are not just genes that have been turned off for possible later turning on. They are past mistakes that we managed to survive. Why would you put such things into a brand new species? If, on the other hand, you were modifying an existing species, they would already be there.

Cheaters on tests are most often caught because they copy wrong answers as well as right answers. Plagiarists are similarly found out because their allegedly original works contain the same misspellings or grammatical flaws as another work. These kinds of identical mistakes are too improbable to have happened except by copying. But evolution is not proud, or original. Plagiarism is its modus operandi.

Unintelligent Design

Many people who struggle with the concept of evolution have no problem understanding the principle of natural *selection*. If an organism manages to develop a new trait that gives it a clear functional advantage, such as eyes, or wings, or arms and legs, it makes sense that nature will select this trait. The organism will do better than its neighbors who do not have the trait and thus pass on more progeny with the trait. But folks often struggle with the part about how those wonderfully adaptive traits manage to emerge to begin with. Eyes and limbs don't just appear suddenly (except in metamorphosis). If these kinds of things take millions of years to evolve, what selective value do they have in the meantime? What good is half an eye, or half an arm? If the selective value does not accrue until the end, what keeps the organism going in the direction of that final goal through lots of useless intermediate forms? Selection can be dumb, but it seems like the generation of selectable candidates has to be intelligent. It has to have some foresight into the final product that keeps it going in the right direction while the product is still not completely functional. Doesn't it?

Another key to understanding evolution is learning to see how the *generation* of selectable candidates is just as dumb as the final selection. For a single mutation to hang around in subsequent generations of an organism, it must have some positive selective value, or at least have no negative value. Neutral mutations are always in danger of being corrupted by re-mutation, since no one will miss them. So for evolution to follow some long-term trajectory toward an eventual complex trait, there must be a continuous gradient of positive selection for (almost) every point mutation along the way. Evolution cannot foresee the end point. It can only blindly follow the selection trail. So what about eyes and limbs? Well, let's look at eyes.

The overall selective advantage of eyes of any kind (or one eye, or four eyes) is to be able to sense light. Up on the surface of Earth, so much electromagnetic energy from the Sun, in what we call the "visible light" range of the spectrum, bounces off ambient objects in a variety of wavelengths (colors), that an organism which can discriminate these wavelengths knows a lot about ambient things, good and bad, in its environment well before it runs into them. This is an advantage. Deep in the oceans, where life began and sunlight is largely filtered out, the colors aren't as important as just knowing the presence of some light at all. It tells you which direction is up, and thus something about the food and prey species that are above you and below you, and it allows you to track the change from day to night which you can use to drive your circadian rhythms. So any mutation on your surface that reacts to even a few photons is useful. A particular class of proteins, the *opsins*, will do this for you. So a simple little copying error in your DNA that misspells some amino acid, which modifies an existing protein into an opsin, will get you started. There are still creatures around today that have nothing more complicated than one of these small, photosensitive spots on their surface. Any subsequent mutation that enlarges the spot by adding more photon-sensitive elements captures more photons. This is a little better. And so the gradient begins.

This kind of eyespot can be found even in single-celled creatures, such as the *euglena*, where it is just a small collection of molecules in the surface of the cell. The same functional design is repeated in multicelled creatures where a collection of specialized cells on their surface forms a much larger photoreceptor that still only senses the presence and intensity of light. The next phase on our adaptive gradient comes from mutations that make the eyespot surface just a little curved inward. A flat receptor surface can't tell you anything about the direction of light. But a curved surface, even a slightly curved surface, gives you a different reading among the photoreceptors on the surface as the source of the light changes. This is good. You could use more of that. The more curved your eyespot becomes, the better you are at discriminating directional changes in things. There are existing creatures, such as the *planarium* (flatworm), that have these "cup" eyes. The optimum of this curvature comes when you have achieved almost a complete sphere with just a pinhole opening. The hole projects a thin beam of photons onto the concave surface at the back of the photoreceptor sphere, which is now functionally a retina. Another existing creature, the *nautilus*, has this kind of eye.

The last phase of the gradient starts when some fortunate mutation deposits a little bit of transparent biomaterial over the pinhole. This is the beginning of a lens. It is harmless in the beginning because it is transparent. It might even have non-vision advantages such as isolating the inner volume of the pinhole eye from environmental contaminants and harmful bacteria. But as subsequent mutations refine it into more of a refracting shape, visual acuity increases. The better the lens, the more adaptive. So selection takes it in that direction. Before you know it, you have a modern eye.

Evolution figured this out without any foresight (excuse the pun) or imagination at all. If you were stumped by the "half an eye" argument above before you read this, you also lacked the imagination to design this eye from scratch. This is not because you are not intelligent, or imaginative, or insightful. Even Darwin lacked the imagination to sketch out this story (though he was confident someone would do it someday). Modern scientists do it with ease. What is it that distinguishes their powers of imagination from yours and Darwin's? They had a lot of extant examples and intermediate forms to work with. You and Darwin did not. They could imagine small adjustments to things they were already familiar with. This is what imagination and foresight *are*, a repurposing of ideas and solutions and patterns you have already experienced. You bend them a little, try to imagine them in a slightly different context, put things together in novel ways, think outside the box a little. We tend to think of imagination and foresight like we are prone to think of life (sometimes) – as an inscrutable flash of something from the outside that magically takes us over some large boundary in one atomic step. We even call it a flash (of insight), a eureka moment, a light bulb in our heads that suddenly turns on. But if you reflect on this phenomenon for a moment, you know you don't go suddenly from a blank mind to a fully formed solution. You were already thinking about the problem, and other near solutions that don't work, when suddenly you see a new connection that enables you to reuse familiar things in a novel way. Insight comes in small increments, leveraging what was already there.

Sounding more like how evolution works? Now let's take it all the way down to dumb chance. In the artificial intelligence field, we study how problem solving works in general, with an eye (there's that eye again) to understanding how it works in humans, as well as making machines that do it better than humans in certain specialized tasks. The formal model of problem solving is that you have some criterion for what counts as a solution that can be mindlessly applied as a test to any proposed candidate solution. For example, let's suppose the goal is to create palindromes (words that are spelled exactly the same forward or backward, such as 'madam'). Then you define the total space of possible solutions (which can be infinite) – in this case, all of the (infinitely many) character strings you can form out of our 26-character alphabet. One strategy that always works, involving no apparent intelligence whatever, is simply to exhaustively enumerate the problem space, one candidate at a time, and check each one against the solution test. Sooner or later, you are bound to find a solution if one exists at all. We call this *brute force* problem solving – trial and error. There is no intelligence in either the generation end or the selection end. Sound even more like evolution?

The brute force strategy can take a long time, so it's often not practical. Instead, we try to inject some human intelligence into the generation process in the form of algorithms that can enumerate the search space more efficiently, learning from near misses how to prune whole branches of it as infeasible. But from a computing perspective, this is just efficiency – speed. The dumb method will also get you there, just a lot slower. Problems that are infeasible to solve by brute force with the computing power of one generation, though, are often solved in the blink of an eye by that same method in the next. Consider, for instance, the recent triumph of IBM's massively parallel Watson computer against the reigning human champions at *Jeopardy!* The Watson program is not all brute force, but its success is based largely on its ability to exhaustively explore a much larger space of possible solutions than Ken Jennings (the reigning human superstar) has in his head. It sure looks intelligent! But if you take the Watson program apart, decomposing it all the way down to the simple addings and subtractings, and shiftings of its binary machine language, you won't find any sudden flashes of insight.

Now we are in a better position to understand how evolution "designs" things without knowing what it is looking for. It is always trying to solve the same, very incremental problem: will the slightly modified thing I just made survive and prosper better than the thing I modified? It is the accumulation of these slight improvements over time that takes it in the direction of complex solutions. It is much, much slower than Watson, but it is exploring a much more vast solution space, and it is exploring that space in parallel on a scale that eclipses Watson's parallelism by many orders of magnitude.

Evolution's solution space has become more coarse-grained, and complicated, and varied over time as the natural selection part gets to choose thumbs up or thumbs down on increasingly sophisticated whole organisms. But its enumeration space, the space in which it generates its mindless, incremental candidates, is still the same, rudimentary spelling workshop that it began with. From an anthropomorphic perspective, we could say that the enumeration element is really just trying to faithfully copy existing DNA spellings, but it makes a few, infrequent errors (we all understand this). It doesn't mean to enumerate alternate spellings. Mistakes happen. But it's not even "trying" to do that. DNA just does this as part of its chemical behavior. It mostly replicates itself flawlessly with just occasional errors. These "errors" cause dividing cells to mindlessly explore the simple space of alternate DNA spellings. It is because these spellings have come to be linked through a cascade of protein and RNA reactions, and gears, and switches, and timings, and signalings to whole multi-cellular organisms, that the selection space has come to be so much bigger and more interesting. It is this intervening linkage of conserved solutions to past problems that allows the ultimate selection space to expand while the original enumeration space remains the same. This is leverage.

Leverage

Leverage is the process by which evolution links DNA spelling errors to traits of the whole organism, allowing a simple mutation to cascade into a complex feature. Genes were first hypothesized as the units of inheritance for visible traits, such as eye color, so there is a long history of thinking of individual genes corresponding to individual traits. When genes were finally identified with DNA sequences that produce individual proteins, the target of gene expression changed, but the thinking did not. Researchers (and the media) started looking for the gene for this and the gene for that. One-to-one linkage between a gene and a trait turned out to be pretty rare. Even eye color doesn't work like this. Genes are linked one-to-one with proteins. What that protein goes on to do in the tangled thicket of biochemical pathways is extremely varied. Some of these proteins become metabolic catalysts, some become promoters or inhibitors of other genes and other proteins, some become parts of bigger molecular machines. Changing one gene often sets off a myriad of effects, some beneficial, some harmful, some neutral. Pharmaceutical companies have come to learn this in spades. Identifying a molecule that has a certain desired effect in a lab dish tells you very little about the enormous cascade of side effects it will have in the whole organism. Many molecules that will cure the specific disease will also kill the patient in other ways.

We have also come to learn that DNA codes for more than just proteins. Only about 2% of human DNA does. The rest (ironically called 'non-coding DNA') can in some places code for sequences of RNA that do useful catalytic and regulatory tasks in the big thicket. And there's more. The emerging field of epigenetics studies how DNA expression can be modified by the state of environmental factors in a particular cell, binding sections of the DNA with methyl groups and wrapping other stretches of it around histone proteins. These factors not only serve to modify which genes will get expressed, but the modification state can be inherited during cell reproduction. Over the eons, these various switching gadgets have evolved to control whole complexes of other genes, so turning off or turning on (or mutating) a single gene can express or suppress an entire complex trait, or the timing or order in which those traits develop. This is different than one mutation creating an entirely *new* complex trait. That can't happen. But if an existing complex trait is built in such a way that small adjustments can result in interestingly different, viable traits, a single mutation can go a long way. The evolution of this intervening machinery has caused evolution *itself* to evolve. We have evolved better ways of evolving, if you will. Some of this machinery is surprisingly modular and robust, so instead of mutations to the switches causing everything to break, you often get viable new combinations of the modules that still work (mostly), but in different ways. This allows the evolution of complexity to accelerate, producing more variety, in less time, in later years.

Perhaps the most striking example of this kind of leverage is found in the *Hox* genes. First identified in fruit flies, these are a series of genes arranged next to each other, in sequence, on the same chromosome. Each codes for a switching protein that controls the development of a particular body segment in the final fly. Curiously, the genes are arranged in the same order on the chromosome, front to back, as the segments they produce in the fly body. The segment of genes is literally a *map* of the fly body. And they are robust. If you switch their order on the chromosome, you get a correspondingly strange fly with switched body segments. If you put the leg segment gene where the antenna segment belongs, you get a final fly with legs growing out of its head where the antennae should be. It was later discovered that all bilaterians (animals with symmetrical right and left sides), from humans right down to flatworms and sea urchins, have this same set of genes for body segments. The proteins in these other animals are not identical to those in the fly, but they are *homologous* – they perform the same function. And they are *robustly* homologous. If you transplant the mouse gene for an eye segment into a fruit fly in the proper place, you get a normal fruit fly eye, not a mouse eye.

There is no particular use for a fly with legs coming out of its head, but you can see the extraordinary range of selection space that this conveniently arranged generation space can explore with the intervening Hox switches. There likely *is* a use for an organism with one more set of legs ahead of or behind the previous set, or more tail segments, or extra wings. Besides simple spelling errors, DNA mutations can also be produced by the deletion of a gene, or duplication of a gene, during copying. In fact, it is gene duplication that provides the source for most new genes. If your gene for protein A gets misspelled producing a new protein B, which may or may not be useful, you may not survive without your lost source of A. So the fortuitous invention of B dies with you. But if your A gene gets *duplicated* instead of misspelled, you simply have an additional source for A. Now if one of your copies of A gets later mutated to B, you and it will survive. Each time this happens in the Hox genes, though, you lose a whole segment or gain a whole segment in the final animal. When you look at the tremendous variety of appendages and body segments in present day invertebrates (such as insects, lobsters, scorpions, centipedes) and the even greater variety of extinct multi-segmented animals from the Cambrian explosion, you see just some of the final animals you can get by deleting, duplicating or misspelling a single gene.

Another way in which small changes to the right DNA can lead to large leaps in final trait space is when the genes that control the *timing* of embryonic development mutate. If genes that control the development of the head and face, for instance, are slowed down relative to the genes for the rest of the body, you can get a final animal with *paedomorphic* traits. The adult form looks like the juvenile form. The effects of this can be seen dramatically in the huge variety of our modern dog breeds. Dogs are only very recently evolved from wolves by their co-adaptation with humans. Humans unconsciously selected for tamer varieties of canines by favoring those willing to interact with them. Much more recently, the various dog breeds we have today were actively selected and bred by humans for particular body and behavioral characteristics. Look at the dramatic difference, for instance, between the juvenile-looking face of a pug and the mature-looking face of a German shepherd. Yet all of these breeds are members of the same species. Their radiated evolution has occurred, with human assistance, only over the last few hundred years. That's an incredible timespan for such morphological variety to arise, and speaks to the leverage you can get by changing the right genes.

Coevolution

Another bit of leverage that natural selection has, to create complex things in short time frames, is *coevolution*. The simple evolutionary model of one organism begetting a slightly modified one requires the conserved internal machinery between certain high-leverage genes and traits to get much of anything interesting in a few generations. Coevolution happens when two or more *different* organisms, with *different* (possibly radically different) genomes, fortuitously discover some benefit from doing things together. We looked at a few of these relationships in the previous chapter. In this case, each organism in the collaborative venture becomes a part of the other organism's environment. Adaptive fitness then means making changes to your genome that track changes in your partners' genomes. When this happens, new traits of the combined, communal organism emerge that are not traits of any member individually. These can be so beneficial to the group, in some cases, that the partners evolve to a point where they can no longer survive as individuals. They have become a collective, super-organism.

This is a different phenomenon than multicelled plants and animals. Multicelled organisms have a single genome, shared by all member cells, that flirts with mutation only once when the designated, single-celled reproducing agents are made (and perhaps twice when sexual versions of these combine). After that, the collective behavior, of all the clones working together as members of the same body, is only changed if the point mutation is selectively expressed or silenced in just the right cell types *and* the places where it is expressed make for acceptable community behavior. This is necessarily a slow process. Coevolution can be faster both because of the initial, emergent kick you get from the discovery of the original partnership (no mutations needed), and because subsequent changes are being made in parallel in independent genomes. Because the members have a long prior history of surviving solo, they can still survive mutations that don't work well together. Multicelled organisms, on the other hand, are toast if the mutation disturbs the communal order. But multicellularity has compensating advantages. Good point mutations are more likely to occur than fortuitous discoveries of partnerships, and the fatality of anti-communal mutations in multicells culls them out of the population quickly.

Coevolution is everywhere in nature, producing inter-dependent combinations of species that we sometimes mistake for single organisms. In most larger animals, a lot of the digestive functions are subcontracted to resident bacteria that break down and synthesize essential molecules that the host cannot, as we saw in the last chapter. Plants, which we depend on to fix nitrogen from the soil, are actually incapable of doing that on their own. Those that do have small nodules in their roots which are actually little homes for colonies of bacteria who do the dirty work (couldn't resist the pun). Lichens are a ubiquitous growth found all over the world, covering an estimated 6% of the Earth's land surface. You may be familiar with them as those grey-green, textured patterns on rocks, and gravestones, and roofs. They also grow in arctic tundra, in deserts, on rocky coasts, in rain forests, in woodlands, and on exposed patches of soil. They were once thought to be a single organism, but turned out to be an elaborate partnership between a fungus and either an alga or a cyanobacterium. The partners come from either two separate domains of life (fungi, bacteria), or two separate kingdoms (fungi, plant). The two partner species are still able to lead separate lives, but together they exhibit behaviors that neither of them does separately.

Flowering plants have evolved elaborate partnerships with animals in which they subcontract pollination and distribution of their seeds, making up for their lack of mobility by hitching rides on the movements of animals. Many insect/plant combinations, where the plants have evolved elaborate color schemes in their blossoms to guide insects to the pollination point, are so tight that neither partner can survive without the other. Fruiting plants have evolved these elaborate, fleshy, high-sugar wrappers for their seeds so that animals will be incented to eat them, then deposit the seeds with their excrement for germination at some distance from the parent. A clever way to solve the inherently territorial problems of looking for sexual partners and better habitats when you have no legs.

There are even instances where one of the co-evolvants is an organism and the other is not. Humans, for instance, have coevolved with language. Languages are an external artifact that gets passed from one generation of humans to the next, mutating and changing as the speakers experiment with new forms. This evolution of ever more expressive syntax and semantics occurred in tandem with humans whose brains evolved not only to process the language but to expect some sort of language to be spoken in the environment of a newborn. The very early brain tunes its synapses for recognizing spoken sounds to the ones it hears in the first six months or so. It strengthens the ones that are repeated and prunes the ones that are absent.

A variation on this theme is when members of the same species develop a gene that supports a beneficial communal behavior of a whole gang of them. These species become social animals, thriving better in groups, or herds, or prides, or tribes. Predator animals, such as wolves, have evolved to hunt in packs, and prey animals, such as sheep, have evolved to survive prey animals in herds. Neither does very well solo. Strictly speaking, the communal evolution within a species is not coevolution, because all the members have the same generalized genome, though often the evolution of highly specialized predator/prey relationships between species is.

Evolving Up and Evolving Out

Perhaps the most famous instance of coevolution can be found in the "single" cell that multicelled organisms use as their basic building block. Nature has given us only two basic designs for the cells that are used to construct this grand edifice of life on Earth, and their very different evolutionary strategies, and the partnerships between them, account for how a lot of the biology on this planet works. The *prokaryotic* cell (pre nucleus), which is truly single, is essentially a small bag of chemicals, with just a single ring of DNA mixed in with all the other protein and RNA catalysts that bang around in the soup. The much larger *eukaryotic* cell (true nucleus), on the other hand, resembles a small city, with the DNA organized neatly into chromosomes inside a separate lipid vesicle (the nucleus), itself resembling a cell, and layers of folded lipid membranes (the *endoplasmic reticulum*) in which more lipid bound things resembling cells (the *organelles*) run around with the various pieces of molecular machinery. In animal eukaryotes, the big organelles are *mitochondria*, which carry out the reactions that produce the rest of the cell's working energy supply. Plant eukaryotes have both mitochondria and *plastids*, which have a number of sub-forms, the most common of which is the *chloroplast* that is responsible for photosynthesis. It is only recently that we've come to realize that these two classes of "organelles" are actually cells in their own right. They contain their own DNA, and reproduce and evolve independently inside the environment of the bigger lipid city. They are the descendants of once free-living prokaryotes, probably bacteria, who somehow got together inside the bounds of another cell and formed a permanent partnership. Since they can get some of their vital proteins from the common host DNA in the nucleus, some of their own genes for these have atrophied over the eons, and they are now unable to survive in the wild.

You may have heard the term "mitochondrial DNA" in accounts about tracing ancestral lineage through genome sequencing. This is because the genomic *identity* of a eukaryotic cell is really a composite of its nuclear DNA (the identity of the mayor of the city) and the DNA of its many mitochondria (the identity of its citizens). Tracing the mitochondrial DNA instead of the nuclear DNA has two advantages. First, since the mitochondria are swimming around in the cell cytoplasm, not the nucleus, they form a part of the female ovum cell, but not the male sperm cell. When the nuclear DNA of sperm and ovum combine, you get a new nuclear identity (a new mayor), but the population of mitochondria in the dividing ovum is still just from the mother. By following the mitochondrial DNA trail, you get a simpler lineage from mother to mother that is not complicated by sexual recombination (you follow the evolution of the city itself rather than its many leaders). Second, since there are many mitochondria in a cell, but only one nucleus, there are many redundant copies of the mitochondrial DNA to work with. Genome sequencing works by first chopping up a cell's DNA and then trying to put the pieces back together with reactants. You have only one chance to get this right with the nuclear DNA, but many with the mitochondrial DNA.

So this "single cell" is really a symbiotic, coevolved community of smaller cells. Prokaryotes were nature's original cell design, and it then took approximately 2 billion more years of evolution to put them together into the eukaryotic city of cells. Eukaryotic cells are the only ones to have subsequently gone on to achieve multicellularity. This is likely due to the very different evolutionary strategies that the two cell types pursue.

Besides the dedicated nuclear membrane that isolates their DNA from the harmful oxidative effect of cell metabolism, eukaryotes also have a lot of evolved molecular machinery for assuring accurate copying of their DNA, for detecting copying errors and damage, and for repairing these. The DNA is the family jewels, and the cell is very concerned about getting their duplication right. This makes the reproductive strategy slow and conservative. Better to have mutations introduce just a little variety, now and then, than to risk losing solutions we already have. This makes sense if these cells are going to be used to build up a complicated, multicelled body. The DNA has to encode an elaborate choreography whereby the right things grow in the just the right orders and at just the right times. The subsequently different cell types must express or suppress just the right genes at just the right times, so that elaborate inter-cellular processes will work as one harmonious whole. A small mistake here or there could break the whole thing. Also, the risk of mutation is borne not just once when the reproductive cells (sperm and ova) are first made, but every time the body cells divide to build the big house. This type of division can happen trillions upon trillions of times in one organism. Beneficial mutations here buy you nothing because only the original germline cells (sperm and ova) get to pass on their genes. But any of a number of mutations that could happen when the body cells divide, that throw off the collective timing of when subsequent cells do and don't divide, and at what rate, can cause cancer – the ultimate breakdown of the entire civilization that can be caused by just a few misinformed renegades. So mutation is generally not your friend. You need to keep a lid on it.

Prokaryotes, by contrast, favor the "anything goes" reproductive strategy. Their DNA is not protected from incidental damage. They make frequent copying errors. They routinely swap whole sections of their DNA with neighboring cells. They also reproduce very fast, sometimes every 20 minutes or so. This leads to a lot of mistakes, and thus to a lot of individual deaths, but it also leads to a lot of innovation. If you make enough copies fast enough, with enough variety in the copies, *somebody* from the original family is likely to survive any predator, or toxin, or environmental change that could possibly arise. This is how bacteria, for instance, evolve antibiotic resistance. With this much instability, it makes sense that prokaryotes cannot support building up the elaborately complex, multicellular cathedrals that eukaryotes can. So instead of slowly evolving *up* in complexity space in a few favored environments, they rapidly evolve *out*, saturating the variety in habitat space, populating every nook and cranny of the Earth with simple, single-celled metabolic solutions to every conceivable set of chemical circumstances. They don't have stunning biological monuments to show for their efforts, but in exchange, they survive the periodic mass extinctions that fell those overly specialized eukaryotic structures when the Earth suddenly changes.

It is because of these two different evolutionary strategies, and their correspondingly different outcomes, that prokaryotes dominate both the organism population and the genetic variety of life on Earth. They are everywhere. We eukaryotes, by contrast, have drastically fewer, but more complex footholds in places where there is liquid water, oxygen, not too much heat, not too much cold, not too much salt, and not too much acid. Everything else we consider "toxic." But there is apparently no such thing as a universal toxin in bacteria land. Whatever it is, some bacterium, somewhere will tolerate it, or even thrive on it. This is why all eukaryotes subcontract much of their basic metabolism to the little critters. We need them. They don't necessarily need us.

One disadvantage of eukaryotes being so conservative about mutations is that it is hard to compete with the faster evolving prokaryotes in times of war. Not all bacteria are our friends. When a species of ours does battle with a species of theirs, we can rarely eliminate all of them because they have so much variety and they evolve so fast. What kills some will not kill them all. They have a much greater chance of eliminating all of us because we are all too much alike. It is thought that sexual reproduction in eukaryotes originally evolved to create this much-needed parity in genetic variety. In true eukaryote fashion, even the random reshuffling of genes in sexual recombination is a properly conservative engine of variety. The genomes of any two individuals of the same species have approximately the same genes, but there is interesting sub-variety in these genes, called *alleles* – alternate spellings, often just single letter differences in the ACGT alphabet, that give the genes slightly different effects. Sexual reproduction gives you a novel mix of the alleles from both parents by first shuffling them a bit when the germline cells (sperm and ova) are created, and reshuffling them again when they are combined to produce the new offspring. The rabid reshuffling occurs within the boundaries of existing genes so it doesn't risk breaking the finely tuned multi-cell plan. But it allows you to express a lot of new variety in the allele combinations in every generation.

***

This has been a long chapter, but it befits the utterly fundamental role that evolution has come to play in our modern concepts of life, and it forms the background material for much of what comes next. If you are among the 97 million skeptics out there, perhaps you can come to see evolution not as a political position that you take in cultural and religious wars, but as a normal slice of science that was universally adopted some time ago. It is no more or less incompatible with there being Gods than is gravity, or solar systems, or the speed of light, or the radioactive decay of carbon, uranium, lead, potassium, and argon. Its observable effects are all around us, and no more magical than cell phones or the Internet.

In our day-to-day lives, we are not accustomed to thinking about ourselves as evolved creatures, any more than we think about baseballs as atomic structures, but the biological gestalt shift hits much closer to home than the one in physics. Baseballs are not composed of smaller and smaller baseballs all the way down to atoms. Physical reduction has thresholds and phase shifts, discontinuities where properties suddenly disappear or emerge because of energy thresholds or the effects of size on our sense organs. But living things are composed of living things, all the way down. There are no phase shifts. So many of the properties we assign to living things, such as intelligence or emotions, do not have clear boundaries where they begin and end, when considered from the evolutionary perspective. Everything is related to everything else. The properties fade in and fade out, and this smooth continuity can play havoc with some of the more discrete concepts that we drag along with us from our simpler past. These continuities affect you even if you feel obliged to deny historical evolution, because you still have to deal with the smooth continuities of embryonic development in the present. If you are willing to take on the full perspective of the gestalt switch, and see yourself as something related to ever so slightly different, earlier versions of yourself from the recent past, as well as earlier versions of the plants and animals you coevolved with, you can make more sense out of why you are the way you are now.

We have, for instance, an irresistible taste for sugar and fat because these are both foods with high energy density. They are high payoff targets if you are foraging for calories in the wild to keep you alive. Humans, in particular, have a similar attraction to the smell and taste of cooked meat because we coevolved with hunting and cooking to get the extra energy needed to support bigger brains. We evolved these tastes in an environment in which the natural sources of these calories were only available now and then, in things like honey and fruit and other animals. So we evolved a neural reward system that incents us to favor them when we find them. The problem now, of course, is that we find them all too readily in the modern environment, but our appetites are still tuned to the old frequency. This is also why we are so susceptible to "supersizing" in fast food. We evolved in a feast or famine environment when the feasts were periodic events against a more constant background of famine. We are incented to keep eating as long as there is still some food in front of us. In the past, this made sense, but in the present, it is all feast all of the time. So we are at risk for becoming obese.

We have a great affection for dogs because dogs coevolved with us from wolves to become our special social partners. They make eye contact with us. They read our emotions to some extent. They can follow our pointing gestures and know that we are trying to lead them to something. Chimps don't do this. We didn't coevolve with them. There is a greater frequency of allergies in developed countries than in underdeveloped countries, and in cities than on farms. This is because our immune systems were evolved to tune themselves early in childhood to the level of pathogens found in the surrounding environment. With our modern urban "hygiene," children no longer get this tuning, so their immune systems remain hyper into adolescence and adulthood, and react to harmless antigens.

This is useful stuff to know. And besides, it's interesting.

5 | Could Life Have Been Different?

I may not have gone where I intended to go, but I think I have ended up where I needed to be.

– Douglas Adams, *The Long Dark Tea-Time of the Soul*

When you think about life from the biological perspective, and the fact that we can't seem to find any compadres on our planetary neighbors, one of the aspects that strikes you is the possibility that life on Earth may never have happened at all. It's apparently nowhere near as probable as the formation of stars and planets. Along with this comes the more playful thought that, since life is contingent on the way certain planetary processes come together, and continues to evolve one way rather than another based on these same environmental buffetings, perhaps we could have turned out very differently than we actually did. Are there perhaps just a few trajectories along which intelligent life *could* have evolved, so any outcome that included intelligent life would look something like us, or are there more degrees of freedom, allowing us possibly to have looked very different than the way we are now?

The Abrahamic religions (Judaism, Christianity, Islam) say that one distinguished God created us in his own image. This would render the way we actually turned out quite unique, suggesting that we are the pinnacle of intelligent life in the universe. Any other variation, on any other planet, would fall short of this ideal archetype. The holy books don't have much to say about aliens from outer space (except for Scientology), since they didn't anticipate this kind of cosmology, so the proper religious treatment of this topic is pretty much unconstrained. This one God could have made different, lesser versions of intelligent life on other planets if he so chose. He is reported to favor humans, after all, and gave them dominion over chimps and bonobos on this planet (well, this is implied). Also, who are we to say that our getting the quasi-divine form was predetermined? This one God could have created anything he wanted. He could have decided to save the best template for a different planet, and made us in some other image. The problem here, of course, is that we have *too* many degrees of freedom. God is not constrained by natural laws, so there's no such thing as one outcome being more probable than any other. If, on the other hand, this God decided to wind up the evolutionary machinery and let it run according to the natural laws (he designed) for this universe, then we can take religion out of it altogether. Could life have been different according to these laws?

Intellectual gadflies from Voltaire and Rousseau down to Mark Twain have suggested that the archetype went in the other direction – that humans created Gods in their own image. Non-European religions had more imagination. Their Gods often take on animal forms, or mixed animal/human forms (the elephant head of Ganesha), or interesting variations on human forms (the many arms of Hindu deities). Popular imagination about other forms of intelligent life didn't really blossom until we got to science fiction. You have to have science before you can have science fiction. But early 20th century science fiction got stuck in a rut of mostly thin, bald aliens with big eyes and enormous heads (to house the outsized brains, presumably) – still pretty close to humans. It wasn't until the George Lucas *Star Wars* films that we really began to exploit some of the available biological variety and imagine intelligence housed in significant departures from the human form. Could we possibly have turned out like one of these?

What Are the Chances?

If there were plenty of Earth-like planets in the universe to provide similar starting conditions, what are the chances that any of them would develop Earth-style life? How probable is our flavor of biology? Given that lipids, nucleic acids, and amino acids have all the right chemical properties for self-assembling into primitive cells, it seems likely that any planet whose physical chemistry can support this kind of biochemistry – or some other "bio" chemistry with similar properties – would develop prokaryote-style life. If our solar system is any model, most planets don't have this (or can't keep this). But there are so vastly many planets in the universe that there must be plenty of targets. The prokaryote design is pretty minimal and appears to be optimal for evolving into almost any conceivable environmental niche, and surviving 3.5 billion years' worth of significant planetary change. So they seem likely, given the right starting conditions. This, of course, is the kind of life that astrobiologists still harbor some hope of finding in small niches on our neighboring planets or their moons. But it takes all the air out of the balloon for the general public. Scientists discover alien life! You mean that's all it is?

Eukaryotes, on the other hand, seem very unlikely, in two respects. First, their particular design seems very ad hoc. This particular combination of parts in this particular way surely must be due to accidental circumstances. There appears to be nothing general or essential about it. We wouldn't expect to find this on another planet, or even on this planet if even a few things had been different. And there aren't any competing eukaryote designs here on Earth, suggesting that this was a one-time occurrence. All eukaryotes derive from a common ancestor. Second, the eukaryotes' unique contribution to evolution appears to be providing the biobricks for multicellularity. Prokaryotes were doing just fine in the single-celled niche. Eukaryotes certainly haven't displaced them down there. Eukaryotes' innovation, which was essential for multicellularity, appears to be the more conservative reproduction strategy that enforces the discipline needed to get cells to cooperate in a complex body. Why would this have initially been an advantage in a single-celled world? We don't really know. It must have been hard to do because nature took about 2 billion years to get there. So it seems that multi-cells are not inevitable, just because you get to single cells. There could easily have been life on Earth, but still not us, or even plants.

On the other hand, eukaryotes could turn out to be like eyes. Eyes turned out to be very likely, evolving independently some 50-100 times (that we know of so far). We lack the current imagination to see the selection gradient for all of the error checking and correcting machinery of eukaryotes because nature erased 2 billion years' worth of intermediate forms. Eukaryotes seem improbable now mainly because we just don't know what would have made them more probable. That may change someday.

Growing Out and Growing In

Once eukaryotes went on to pursue multicellular evolution, they went down three separate paths, or at least three separate *surviving* paths: fungi, plants, and animals. Why three, and not two or five? We don't know. Would multicelled life on another planet have gone down these three paths? We don't know. You may be surprised that fungi got a whole kingdom of their own. They don't seem very impressive. But that's because you don't know fungi very well. You may be familiar with them as that furry white stuff on spoiled food, or that black stuff in your shower, but these are just a few of their many body forms – and it's very common for fungi to transition into several different body forms in the course of a single life cycle. They can reproduce by budding, or sporification, or single-celled division, or sexual recombination. They have it all. They can be microscopic, as in the yeasts that inflate our bread or ferment our beer, or have large, elaborate fruiting bodies such as mushrooms. A single individual can extend for miles on the forest floor, sending out branches in all directions which face the unique challenge of determining whether the other biological things it encounters are friends, foes, or other parts of itself.

We, of course, eventually emerged on the third branch. Could something as intelligent as us have emerged on either of the other two? Could we have been intelligent fungi, or intelligent plants instead? Probably not. The key reason is that we need something like a nervous system to support intelligence. Animals got these; plants and fungi did not. And this is probably because plants and fungi pursued a similar architecture for embryonic development and growth, where animals eventually pursued a different one. We will develop this argument over the rest of this section, but the short version is this. One architecture is suited to movement, the other to staying in one place. Nervous systems got their start as a solution to the movement problem.

Multicelled fungi and plants grow *out*. They remain centered in a fixed location and extend their territory outward by growing further extensions of themselves until environmental factors like lack of sunlight, or temperature, or other plants stop them. They have an embryonic body plan for a few basic shapes that they achieve rather quickly, then they continue to repeat those shapes, recursively, by embedding a smaller version of the shape in an outward tip of the last one. They don't have a predetermined plan (typically) for when to stop. So there are no juvenile and adult stages like in animals. They just keep growing. Each individual is shaped somewhat differently. The similarity between them is the general similarity of their embedded parts. They are similar to themselves. Their final form reflects characteristics of the environment where they are originally rooted. Their genomes encode a development plan that improvises as it goes to take these factors into account. It's important to conform to your immediate environment if you are going to spend the rest of your life there.

Multicelled animals (once you get past sponges) grow *in*. Their genomes encode a plan for a fixed final shape that every member of their species will achieve. They all look roughly the same when they are done. The embryo is progressively subdivided into more and more specialized compartments with dissimilar shapes and dissimilar functions. Developing cells will migrate at specific stages to reset the shape of developing internal parts. All of this growth takes place *within* the boundary of the initial embryo. It just gets bigger. The genome encodes a plan for when the juvenile form is done and ready to face the environment, and another plan for any continued growth and development to get it to an adult stage. After that, growth stops except to repair and maintain the final form. The development plan does not consult the immediate environment (much). It already knows what to build. It is designed, by prior evolution, to make something capable of surviving in many variations of the local environment. It has this luxury because the final organism is capable of moving. And the many movings of its ancestors have encoded the variable characteristics of many environments in its genome.

The growing-out plan of plants and fungi is clearly simpler, and arguably easier to evolve. You don't have to encode much complex structure, or much about past environments, into your genome. The organism is literally shaped by its environment in each generation. This is probably why the most primitive of animals, the sponges, look like this. They differ from plants because they lack the plastids in their cells to photosynthesize energy from sunlight, but they can still absorb microorganisms into their tissues in the sessile, waving-in-the-water formation of real aquatic plants. It's easy to imagine that the early world of multicelled life was essentially a growing-out world like this. If evolution had stayed on this course exclusively, there's not much chance we would have become intelligent seaweed. What got the growing-in strategy started was the invention of large-scale movement and predation.

Plants do occasionally employ bits of the growing-in strategy, in the generation of flowers and fruits for instance. These are periodic, complex variations in the self-similarity structure that serve specific adaptive purposes. And animals employ bits of the growing-out strategy in specific internal places such as lungs, and blood vessels, and nerves. So it is likely that there was constant experimentation with both architectures in the ways early embryos divided. This eventually gave us the two specialized cell types that support autonomous movement: sensors and effectors. Sensor cells detect some environmental factor such as light, or compression waves, or contact, and change their state accordingly, emitting something to neighboring cells. Effector cells produce local shape shifts in response to some signal. Put them together and you have the primitive basis for reforming your shape, in real-time, in response to changes in your environment. Single-celled creatures had already exploited this kind of movement for adaptive advantage, with sensor proteins on the cell surface and flagella-style tails. So its re-invention in multicelled form was just waiting to happen.

Once you can move around to look for your food, instead of passively waiting for it to float by, you have an advantage over your sessile animal neighbors. Perhaps the reason that plants didn't go in this direction is that their food source, the sun, is omnipresent. You don't have to go looking for it – it comes to you. But if you have to eat other organisms for food, this is a big deal. Eventually you get to the point where these bigger swimming things have mouths capable of ingesting multicelled food, and active predation begins. This changes everything. Now if you are sessile, you are a sitting duck. These moving things will graze on you. If you are a moving thing yourself, you need to be constantly on the offensive and defensive to keep up with your neighbors. You need to get stronger, faster, more discriminating, more complex in your movements. This forces you to continue to evolve further down whichever of the development trajectories you are already on. Natural selection begins to prune away the hybrids.

If you are a plant, the simpler and more robustly repeatable your recursive template is, the better you survive grazing. Since each extension to your overall structure is essentially a smaller version of the same thing, any one of them can serve to regrow a portion that gets grazed away. There is no single point of failure in your structure such that if you lose it you are toast. But you also can't afford to evolve such a single point of failure in the future. There is now a cap on your organizational complexity. You must continue to be a simple structure of repeatable parts. You can only develop a complex structure that is not essentially connected to your biostorm. It has to be something you can afford to lose. So plants turned this principle on its head. Since I'm going to be eaten anyway, why not entice my predators with complex organs (like fruit) that will help spread my seeds around. You have no need for a nervous system that causes you to recoil or warns you with pain. You can't get away anyway.

If you are an animal, you now can't afford to stay too simple. Unless you can find a protected eco-niche somewhere, you have to keep up with the Joneses. You have to eschew any growing-out on your surface that makes you unbalanced. You have to tightly control any internal growing-out so as not to upset the internal clockwork. You are incented to develop more complex internal organs that enable you to better compete. This also causes you to develop single points of failure, like hearts. You are always vulnerable to one big bite in the wrong place. But simplified repeatability is no longer an option. You just have to keep doubling down on these points of failure to get better and better at *avoiding* the big bite, with things like armor, and teeth, and speed, and *intelligence*.

Symmetry

When we think of symmetry, we don't normally think of things from nature. Plants, rocks, coastlines, clouds, storms – these are all irregularly shaped things. Humans, it seems, create most of the symmetrical things in the world artificially. But take a closer look at animals. All of the multicellular animals are either radially symmetrical, like octopuses or jellyfish, or bilaterally symmetrical, like us. We bilaterians have right and left sides that are mirror images of each other. Plants are indeed asymmetrical because their self-similarity is due to recursive embedding. You can't divide them along any set of top-level axes that results in rotationally identical pieces. Is this just a coincidence, or do symmetry and growing-in go together in some essential way? Could we have been asymmetrical intelligent beings? If not, could we have been radially symmetric intelligent beings instead of bilaterians?

There probably is a hard connection between symmetry and movement. Think about it. If you have some sort of appendages that propel you around in the water (and eventually on land), such as tentacles, or fins, or legs, they must be symmetrically balanced for you to move in a straight line. If you have more on one side than the other, or some are shorter than others, you will move in circles. Think of rowing a boat with two oars of different length. You're not going to catch prey that can move along straight lines. And predators that are bearing down on you along straight lines will catch you. Even if you are some sort of worm or eel, without appendages, your shape has to be bilaterally symmetrical for you to undulate in straight lines. So competitive movement strongly selects for symmetry. Creatures that develop it will have an advantage, and creatures that lose it through subsequent mutations will be culled away.

You might think it is difficult for random mutations to develop a trait like symmetry that seems to require some delicate coordination between multiple parts. But, like nature's tendency to form spheres, it is actually more probable than asymmetry if you are pursuing the growing-in development plan. Embryos progress by individual acts of cell division, after which the cells stick together rather than going their separate ways. Unicellular reproduction has turned the division process to favor two daughter cells with the same characteristics. So if multicellular organisms don't intervene to break this sameness when cells divide to form a larger cell mass, you get symmetry for free. You get daughter cells of different kinds only by breaking this default symmetry with gradients of proteins in higher concentrations at one side or the other of the parent cell, giving the cell a polarity. These polarities are the more unlikely properties that require fortuitous mutations. If you don't intervene to break the default symmetry, subsequent divisions on both sides of a first division will go on to develop the same things on both sides.

Radial symmetry was popular with the earliest animal forms, probably because it is simpler, but once bilateral symmetry came along, it dominated the animal development program. The radial animals are shaped like bells with the tentacles facing one direction and the head/body the other. So they have just one axis for orientation and movement: leading and trailing. The earliest forms, like medusae, use this for up and down movement. Their side-to-side movements are really driftings on the currents. The bilateral plan created three axes for orientation and movement, forward/backward, up/down, and right/left. This proved to be more versatile. Topologically, the basic shape is that of a tube (recall the earlier discussion about multicelled bioreactors), but a *tube in context*. It is assumed that the tube will be oriented in parallel to the surface of the Earth, creating a distinguished top-half/bottom-half relative to gravity, and that one end of the tube (head) is the leading end for movement and the other (tail) is the trailing end. Optimizing for movement in this "parallel to gravity" orientation encourages you to vary the front and back ends, and the top and bottom halves, in asymmetrical ways to exploit environmental asymmetries. You want things like mouths and eyes on the leading end facing prey, and things like tails and rudders on the back end to steer you away from predators. But you'd better not mess with right/left symmetry. The things in your environment are pretty much randomly arranged on the right and left, even in the context of chasing something or fleeing from something. So there are no right-left environmental asymmetries to exploit. All you can do is screw up the efficiency of your motion by fiddling with this.

Evolution went on to exploit this basic pattern in countless variations. With the help of leveraged genes like the Hox sequence, it added and subtracted segments, added and subtracted appendages, and modified them into fins and tentacles, and antennae, and claws, and pincers, and arms and legs, all the while keeping the creatures in the original horizontal-to-gravity orientation. So worms, fishes, alligators, and mice are all essentially elaborations of the same template. Eventually, a few creatures, like us and birds, rotated 90 degrees and switched to a vertical-to-gravity orientation. The old front and back became the new up and down, and vice versa. But birds still fly in the horizontal orientation, and we swim that way. It is perfectly feasible that we could have ended up in the horizontal orientation or that we could have had different sorts of appendages. The multi-armed Hindu format would have been easily reachable with a few Hox gene duplications.

A lot of our internal organs – lungs, kidneys, testicles and ovaries, brains – come in two symmetrical parts, but this is inessential. It is a side effect of the overall bilateral segmentation of the embryo. Others, such as the heart and liver, are unnecessarily single. We could have had singular brains and lungs, and multiple hearts and livers. Sense organs that determine the presence of things at a distance, such as eyes and ears, naturally come in pairs because of this same bilateral segmentation, but they are much more effective in pairs anyway, so we are likely to have had *at least* two of these, if not more. With two, you get multiple, slightly offset readings of the incoming energy and this tells you more about where the source of the energy is coming from in three-dimensional space. Noses, although singular, have two chambers to achieve this same effect in determining the source of smells. This same advantage does not extend to mouths or tongues with taste receptors because these operate on contact. You don't taste something or eat something until you are actually touching it, so its approaching direction is irrelevant. So one mouth is likely.

The upshot for symmetry is that because our nervous systems originally evolved to support movement, and efficient movement requires symmetry, we were probably destined to be symmetrical beings. But if evolution had gone on a different path early on, it is conceivable that we could have ended up as radially rather than bilaterally symmetric beings. This would have lifted the constraints and influences of pairs and we could have conceivably been creatures who cartwheeled around the Earth as multi-appendaged spheres with brains in the center and eyes, ears and perhaps even mouths on each appendage.

Nervous Systems

We are intelligent because we evolved nervous systems. Does this mean that any intelligent life must have a nervous system? Could we have evolved intelligence in some other way? To answer these questions, we must first have some minimal notion of what intelligence is. Intelligent creatures can react to their environments in real time. They can sense and move toward food. They can sense and move away from toxins and predators. They can sense others of their kind and engage in cooperative behaviors with them. Bacteria can do all of these things without a nervous system. Plants can do all but the last one, and though their movements are too slow to count as real time, they would be fast enough in a world without grazers. Their roots move toward food and away from toxins. Their photosynthetic elements move toward sunlight. They don't have nervous systems. They move by directional growing. Jellyfish do have nervous systems, but they are not social. Yet "the mind of a bacterium," "the mind of a plant," and "the mind of a jellyfish" all sound like insults rather than descriptions. You need to do more.

Because jellyfish got to this minimal capability by way of nervous systems, future animals got on an evolutionary trajectory toward more complex nervous systems that eventually crossed some threshold that we would consider intelligent. Bacteria and plants topped out and got no further. You might even say that bacteria were further along than plants. They had real time movement and sociality. Multicelled organisms had to reinvent many of the single-celled functions, such as metabolism, reproduction, bodies, movement, signaling, and cooperation by using combinations of cells in place of combinations of molecules. Some of these functions, such as reproduction, really couldn't be improved on, so the single-celled version still forms the basis. Signaling and cooperation went the other way. Doing it with cells instead of molecules allowed the functions to be improved exponentially. So the quick answer is that molecular signaling can only take you so far, making single-celled intelligence infeasible, and plants failed to develop cellular signaling into real time units like neurons. Animals did. There likely is no other biological way to bootstrap your way to intelligence.

The basis for a nervous system was already there in the first animals, even before they gave up the sessile lifestyle. Sponges exhibit a type of purposeful, real time movement. They feed by filtering water through the many small pores on their outer surface and then expelling it out of a large hole at the top. Microorganisms and other nutrients in the water are absorbed by the *choanocyte* cells lining the inner surface. The choanocytes also contribute to the body-wide movement that causes water to flow through, by waving their attached flagella. These cells, and a class of muscle-like cells called *myocytes* that surround the openings of the pores and expand or contract to modulate the diameter of the opening, are known as direct effectors. They are stimulated directly by environmental conditions, mechanical, chemical and thermal, and respond directly with an appropriate motion. This same function in single cells is called *taxis*, whereby an appropriate molecular cascade, internal to the cell, links receptor binding events at its surface to appropriate movements of the cell as a whole. As cooperating members of a larger, multicelled concern, these individual cell movements can sum up to produce meaningful movements of the organism as a whole.

Plants never developed these fast-acting elements (except in rare cases like the Venus flytrap) because their food (sunlight and root nutrients) doesn't move very fast. Because animals did, it was only a small evolutionary step from direct effectors to neurons. Even as a group, each individual direct effector cell senses and acts directly. When they detect something, they do something about it. The net effect of converting some of these cells to neurons is to introduce an element of indirection between stimulus and response. When neurons detect something they don't act, they just talk about it. They sense and report. Their actions are inherently information events. They depend on some other cell or cells downstream to act on what they report. They are sentinels.

The advantage of indirection is that it decouples input from output. If input is connected directly to output, then you largely get the same response every time. If you place a population of reporters between the stimuli and the cells that will ultimately respond to them, you allow for groups of responders to behave differently given different combinations of stimuli. This allows for many different patterns of input to be detected by the same group of sensors, and many different patterns of response to be constructed by the same group of motor cells. The firing pattern of the sensor neurons is a representation of events in the outside world that caused the stimulus. They regularly occur together. This is the real innovation introduced by neurons. They make it possible to sense the outside world without directly acting on it. This indirection allows the appropriate response to be varied over time without having to redesign the sensory system. The sensory and motor systems can evolve somewhat independently, each conserving its own greatest hits without fear of breaking the other. The sensory system responds symbolically and the motor system interprets the symbols.

This symbolic indirection can then be expanded in the middle by the introduction of , cells that sense symbols and respond with further symbolic interpretations, eventually making it possible for an organism to learn – to adjust the intervening interpretations on the fly – and thus to respond more precisely to fine-grained changes in its environment. In jellyfish, sensor neurons are connected to a secondary set of motor neurons, and those neurons, in turn, project into the muscle cells. This two-layered nervous system is laid out as a diffuse net, reflecting the fact that the sensors and muscles are relatively smoothly distributed over a large part of the body. But at more specialized parts of the body, such as the mouth and the base of the tentacles, there are small concentrations, often ring-shaped, of that both receive from and send to other neurons. It is no surprise that these concentrations tend to emerge in areas where the body specializes in some particular task that requires more fine-grained interpretation and response to environmental events.

This, in a nutshell (or a jellyfish body), is the neurological basis for us. Concentrations of are called *ganglia*. These emerge, as in the jellyfish, in areas of the body that need more precise and variable computation because they give you an adaptive advantage. As evolution moved from the radial to the bilateral plan, these ganglia became more concentrated toward the head to be close to the long-range sensors like eyes and important moving parts like the mouth. Brains are just ganglia on steroids. The relatively direct sensor-to-effectors connections of the most primitive bilaterians became our peripheral nervous system, and this system became routed through the central control center of our brain via the long ganglion of the spinal cord.

Our modern brain comes in three sections. The oldest, brainstem section we share with amphibians, like frogs. These parts directly regulate heart rate, blood pressure, breathing rhythm, body temperature, digestion, control of sneezing, coughing, vomiting, and switching between waking and sleeping, all typically below the level of our conscious control. These parts also contain a complete set of hardwired programs for regulating behaviors for feeding, fighting, fleeing, sex, and temperature maintenance (shivering, for instance). The second section, the limbic system, we share with all mammals. It houses our emotions and motivations, like fear and happiness, and the processing of memories. These go together because memories are tagged and retrieved with emotional markers, fear being the most salient. The last section is the cortex, which selectively overrides the other sections based on reasoning, planning, and conscious control. We share a version of the back part of the cortex with lots of our animal neighbors, but we are distinguished by our exponential expansion of the frontal part. Somewhere along this continuum of improvements, intelligence emerges. We had to be animals.

Skeletons

Around 450 million years ago, well after nervous systems evolved, the animals split into one of two fundamental branches: vertebrates and invertebrates – animals with and without a backbone. The names suggest our orientation as vertebrates. The other guys are defined by some feature of ours that they lack, rather than some compensating feature that they have. The invertebrates have a better claim to being the foundational form. Up to a certain stage of embryonic development, both of us look similar. Then as the *notochord* (the precursor of the spinal cord) begins to develop we look like two versions of the same fat worm rotated horizontally by 180 degrees. Their notochord develops along the underside of the body and ours develops along the top surface. One of us is an upside-down version of the other. It's as if some fortuitous event flipped one version of the embryo over (probably us), and both versions worked pretty well, so evolution continued down both paths. We have diverged rather profoundly ever since.

In fairness to our own point of view, vertebrates are more similar to each other than the larger class of invertebrates are to themselves. The key difference is in skeletons. In order to be more than just a

Jello-like mass of cells, your body plan needs to build some kind of hard structure to hold your shape, support your weight, and protect your inner organs. Plants do this in the brick and mortar style by forming rigid cell walls between cells. Animals, which need to be more flexible for movement, do it with large pieces of hard material, like bones or shells, architected to support large collections of more flexible cell tissues. Our distinguishing feature, the backbone, says it all. It is on our *backs*, housing our flipped over spinal cord, and it is a *bone*. Invertebrates don't have bones. For whatever reason, we started anchoring and protecting our top-level central nervous system with bony fragments of calcium. This turned out to be such a good idea that our evolving body plan just kept doing this. Each moving part has a bone (or two) running down its center, and all of the soft tissue grows around the outside of the bone(s). We have *endoskeletons*, bony framing on the inside on which you attach tissue.

Invertebrates took a different approach. They have *exoskeletons*. Each section of moving parts is surrounded on the outside by a hard shell. The inside is soft tissue, all the way in to the middle. They both work. If anything, invertebrates have explored a much richer variety of body plans with exoskeletons. They have all of the six-legged (insects), and eight-legged (spiders), and n-legged (various arthropods like lobsters) varieties. There are vastly more of them than there are of us. One disadvantage of the exoskeleton plan is that you can't grow continuously. Once the outer shell forms, the inner growth has hard limits. To get bigger, you have to molt – shed the outer shell and grow a bigger one. This can leave you in a rather vulnerable, armorless condition during the transition. This is one of the downsides of evolution having no foresight. What looked like a good local solution at first had a big gotcha later down the road. If this had ultimately been a showstopper, natural selection would have pruned the invertebrates away. But it didn't. So it's viable. This limitation may, however, account for why invertebrates don't get very big. Vertebrates have no fixed limits on growth, and we once got as big as the dinosaurs.

Could we have gone down the invertebrate path instead? They have full-fledged nervous systems. They can be very social and intelligent looking, e.g. ants. Modern science fiction uses insect-like creatures more often now as space aliens. But the naturally occurring invertebrates here on Earth seem to have topped out in intelligence compared to mammals and primates. Is there a reason for this? It could be size. There is a strong correlation between degree of intelligence and number of neurons. Look at how much of our relative body mass is taken up by our big brains. Perhaps it is the vertebrates' ability to grow large bodies via the endoskeleton plan that enabled us to ride neuron growth up to the level of human intelligence. The large, bug-like creatures from outer space must have found some solution to the growing problem other than molting.

***

So we were destined to be animals, rather than bacteria, fungi or plants. We would almost certainly have been symmetrical to have gotten nervous systems at all, and most likely not have had exoskeletons to limit our growth to critical brain size. But we could possibly have been radially symmetrical creatures with endoskeletons. On the bilateral side, it's not clear why we couldn't have been intelligent dinosaurs, or intelligent dogs, or intelligent birds. There is nothing obvious that would *prevent* this from happening. On the negative side, we are the only vertebrate species that got this far. There was no convergent evolution, as with eyes. Why is that? On the positive side, we are a very young species. Perhaps we are just the first intelligent species on this planet. There are plenty of other primate species around. If we kill ourselves off, perhaps one of them will get there again.

A last dimension to consider is our continued evolution. Many species have lost features that got them to a certain point, once their habitat changed so that they no longer needed them. Whales, for instance, were once land mammals. They lost their legs and other tetrapod features when they went back to sea. In our case, bilateral movement was necessary as scaffolding to get us to intelligence, but now that we have intelligence, and have developed technology that renders our bodily movements unnecessary, a lot of our moving body parts could now atrophy (it would take millions of years, though). We could become primarily brains, motoring ourselves around on golf carts, feeding and reproducing artificially. We could replace our eyes, ears, noses and mouths with artificial sensors. We would still need something like our vocal cords to generate language, but voice recognition software could encode it into messages on radio waves. We might conceivably be visited by intelligent folks like this from other planets. Or we might visit them someday in this form. The meaning of life would be very different then.

6 | Me and My Bacteria

New Rule: Instead of killing 99.9 percent of germs, Lysol has to just go ahead and kill them all. Why spare the remaining 0.1 percent? So they can return to their villages and tell the other germs, "Dude, do not mess with Lysol"?

— Bill Maher, *The New New Rules: A Funny Look At How Everybody But Me Has Their Head Up Their Ass*

Humankind has only relatively recently become acquainted with microorganisms. They were just too small to see until we invented instruments with which to see them. Prior to that, our acquaintance was only with a few of their observable effects, and these were largely negative effects: disease and the spoiling of food. Spoiled food you could deal with, because you got a clear sign from the creepy colors and foul smells. But infectious diseases were insidious. They passed mysteriously and invisibly from person to person, often without direct contact. Evil spirits. Vapors. Bad air (malaria). Inscrutable spirits filled the void of mechanical explanation, as they often do, but in the case of diseases, this non-explanation was particularly frustrating. The plagues of Europe seemed to fell whomever and whenever, regardless of precautions or interventions. If these were punishments from God, there seemed to be no consistency to the behavior that brought them on. The deadly spirits seemed to take both the good and the bad indiscriminately. If these diseases were actually *sourced* by evil spirits, acting independently of God, why couldn't God intervene, at least to save the righteous?

Humans were observant enough to connect this poisonous vapor, or *miasma*, to the other observable manifestation of microbes – rotting organic matter. It seemed that there was always something fetid, or foul smelling, or decomposing in the vicinity of these outbreaks. This miasmatic theory of epidemics was standard medical lore well into the 19th century. When we finally figured out that both of these phenomena are caused by microscopically small creatures, the negative cast was already set. We had been at war with what we thought were evil spirits. We had actually been at war with another life form: germs. The term 'germ' was originally neutral, referring simply to microorganisms. Anton van Leeuwenhoek used the term in the 17th century to refer to the little "animalcules" he observed under the early microscope. It wasn't until the germ theory of disease that we had any clear idea what these creatures do in the grand scheme of things. Then we found out. They are out to kill us. Better to kill them first.

Given that we still can't see them in everyday life, we have inherited the original phobia concerning the undifferentiated mass of invisible agents out to get us, creeping into crevices, living on things around us, infecting us before we know it. The advertisements for our personal hygiene products are constantly urging us to kill germs – all of them. We are advised to cover our mouths and noses, to wash our vegetables and kitchen counters, to be careful what we touch at the gym and on the subway. Howard Hughes is famous for (among other things) having taken this phobia to such extremes that it destroyed his health. Doctors routinely prescribe broad-spectrum antibiotics as a kind of napalm in this war against the bacteria, under the theory that the only good bacterium is a dead bacterium. The very thought of them crawling on you, getting in your mouth and nose: Yech! Germs!

Well, remember how you are really composed of a large community of cooperating cells that originated by successive divisions from your first cell? You have about 37 trillion of these, according to the most recent estimate. That's pretty big. But that huge bio-community of cells that makes up your body, that jointly works together to enable you to survive and prosper, actually consists of about 137 trillion cells. Yes, that's right. About 75% of them are not yours. They are the individual bodies of your personal bacteria (mostly). They live on your skin, in your hair, in your eyes, in your ears, in your nose, in your mouth, in your stomach, in your intestines, in your vagina. They digest your food and maintain the health of your skin and gut lining (among other things). You coevolved with them over millions of years. You depend on them; they depend on you. You can't survive without them. Germs are us.

Colonization

Over the last 6 years or so, the microbiology community has consistently estimated that the ratio of them to you is closer to 10 to 1. This was based on an estimate of 10 trillion human cells per body. The 37 trillion figure is a very recent (at this writing), more fine-grained estimate for the human side, a counting problem we have been studying for centuries. The bacterial count hasn't been re-estimated yet (stuck at 100 trillion). It's still a shot in the dark. In fact, the whole subject of humans being mostly bacteria, and what that means for our health and wellbeing, is a new area of research. We're still mapping out the territory.

You might be wondering how this is possible. You've probably heard things like "the human body is mostly water." That makes sense because cells, and just about everything between cells, is mostly water. But if we are really three quarters bacteria (or more), how could physiologists have missed that? Well, only 1-3% of your total body mass is taken up by your bacteria. Remember, your cells are eukaryotes; theirs are prokaryotes. Theirs are *much* smaller. They dwarf us in cell count and total genes, but we still dwarf them in pounds per square inch.

So how does something like this happen? How can these vast populations of microbes that make up our bodies coevolve with us? How is the genetic information for all of these extra critters passed on from one human to the next? Is it encoded somehow in the human embryo? No. Though that for a different population of prokaryotes is. Recall the mitochondria that make a living inside our eukaryotic cells. There are, on average, about 100 of them in each human cell. They do indeed evolve on the inside with us as we carry them along in our reproductive cells. If we added them to the bacterial count we would really be dwarfed. But they are really more like our naturalized citizens. They are counted in our census, though we only count the whole household and not the family members. The other bacteria are more like the resident aliens of our vast nation, our guest workers. They work on the outside of our cells, so they can't follow us through reproduction. They have to wait until a whole new, sterile human is born, and then rapidly colonize it from the outside.

This may seem like an impossible feat of choreography, getting this balance just right from one generation to the next, but this is the norm in the world of multicelled creatures. Almost every creature of significant size subcontracts to the bacteria (and other microorganisms) for survival. Big organisms are essentially their environments. They are adapted to the rhythms and timings of these environments. Just as migratory birds and fish and mammals have evolved to follow the sunlight and seasons to different areas of their geography, bacteria follow the human reproductive seasons in their periodic migrations. Starting in the first trimester of pregnancy, the population of bacteria in the mother's vagina begins to change in preparation for the coming change of "season." A particular bacterium, *Lactobacillus johnsonii*, which is normally in the gut where it assists in the digestion of milk, becomes one of the dominant species in the vagina. There is no milk to digest there, but changes in the local "weather" tell the bacteria that soon a whole new version of the environment, with a whole new gut to colonize, will come sliding by.

The human infant is largely bacteria-free just before birth. But on the way out, it is bathed in the resident bacterial communities of its mother's vagina, and to some extent her intestines. Skin to skin contact brings in the residents from this eco-niche of the mother as well. Shortly after birth, an ecology of bacterial species specific to infants stabilizes in the newborn's gut. Breast-feeding plays an essential role in this first colonization, both because the mother's milk contains some 600 kinds of bacteria to start the colony, and because it contains not just food for the baby, but also food for the baby's bacteria! Here we see some of the effects of those millions of years of coevolution. In addition to the common lactose sugar that is the staple food for mammalian infants, human breast milk also contains some more complex sugars called *oligosaccharides*. The human infant lacks the digestive enzymes to metabolize these. So what are they doing there? A particular species of bacteria, *Bifidobacterium infantis*, which is important for the initial colonization, does digest these. It is their favorite food. If you feed them early, they in turn spread out and prevent more harmful bacteria from establishing a foothold. They are also involved in maintaining the health of the epithelium that lines the inner surface of the intestines. We're all in this together.

With the introduction of solid food, the infant's community of gut microbiota changes to match the new assortment, and molecular packaging, of nutrients, and then changes again after weaning when the lactose ends. By about 3 years of age, the toddler's community begins to resemble that of its parents and the process is complete. Because we think of modern humans being born in sterile hospitals, with all kinds of interventions to improve the chances of newborn and mother, we have tended to ignore the very different circumstances in which this delicate ecology of initial colonization evolved. Caesarian sections, for instance, were first introduced as an exceptional, emergency procedure. They have gradually become responsible for about 32.8% of births in the US, often, it is speculated, for the convenience of facilities and physicians. But caesarian birth bypasses the vaginal and intestinal inoculation that we have so carefully evolved, leaving the infant's initial community of gut microbiota looking more like that of the parent's skin. Not coincidentally, C-section infants have higher rates of allergies, asthma, and autoimmune conditions. We have experienced unintended destruction of ecologies like this at the macro level many times, when we drain swamps, or dam rivers, or extinguish all forest fires, or remove keystone species from a food chain. Now we have to be sensitive to how these ecologies work in an entirely new habitat – us.

The Human Microbiome

You have probably heard of the Human Genome Project, the worldwide, collaborative research effort to sequence and map all of the genes encoded in human DNA. It began in 1990 and was finally completed in 2003. We had known about genes in their DNA form since the 1950s. You may not have heard of the Human Microbiome Project. It started much more recently (2008) to make a similar kind of standard map of all of the many little critters living on and in us. We would like to know which species live where, and the nature of their much larger collection of genes that round out what we have come to know as human-hood. But because the phenomenon we are trying to map has only very recently come to light, we are not yet sure exactly what we are looking for, or how exactly to go about looking. Unlike the Human Genome Project, there is no fixed criterion for when you are done.

The first results were published in 2012, involving more than 200 scientists who studied 242 healthy human subjects over the course of two years. The individual genomes of bacteria found in 15 different locations of the body were sequenced, yielding more than 5 million genes (humans have fewer than 20,000 of their own). Our resident microbes, collectively referred to as *microbiota*, consist of more than just bacteria. There are also archaea (another kingdom of prokaryotes), fungi, and some single-celled eukaryotes, but the bacteria dominate. Different populations of these microbes specialize in the different territories of the human geography: skin, mouth, eyes, respiratory system (nose and lungs), stomach, intestines, and vagina. Skin covers a lot of human territory, so the concentration of species varies with the particular location on the body. In general, these species feed off waxy secretions and perspiration of skin cells, and provide the return benefit of a moisturizing film that keeps skin from drying out and cracking. Your perspiration is inherently odorless. It is your bacteria that give it its distinctive smells. Relatively few species are in your lungs because they do more harm than good there, so your respiratory system has mucus, sneezing and coughing to catch them and expel them. The stomach also contains fewer residents because of its hyper-acidic condition. The vast majority of your bacteria are in your intestines, and the vast majority of those are in your lower intestine.

One of the goals of this survey is to establish some baseline for what counts as a "normal" microbiome in a healthy human. It is hoped that this will help identify abnormal microbiota populations that can be associated with disease conditions, and perhaps lead to microbial intervention therapies. A single such reference standard is proving hard to realize because microbiomes vary with geography, culture, and diet. There are substantial differences in what is normal for a resident of a Western industrialized nation and a hunter-gatherer in Africa, for instance. The typical Japanese microbiome contains a microbe that enables humans to digest seaweed, which the rest of us lack. The modern Western diet, high in sugars, fats, and processed foods, as well as the modern Western use of antibiotics and sterile environments, has resulted in substantial evolution of the Western microbiome over a very short timeframe. This brings with it some concern that the diversity of original microbiota species is declining at a rate that will render some of these species extinct. Humans have spent many millions of years getting this symbiotic balance about right. Although a "proper" community of microbes can often be retroactively restored by changing diet and other environmental factors, if some of the members of that proper community are no longer with us, this will not be an option.

Sampling microbiomes from indigenous people in remote parts of the Amazon that have had little contact with modern civilization reveals marked differences with the typical Western profile. These "original" microbiomes exhibit much more diversity of species, including some that have never been sequenced before. There is a growing suspicion that these differences have something to do with Amerindians' significantly lower rates of allergies, asthma, obesity, type 2 diabetes, and cardiovascular disease – all conditions that have been growing disproportionately in Western societies recently.

One bacterium, common in non-Western microbiomes, which is now nearly extinct in Western ones, is *Helicobacter pylori*. It resides in the stomach, and has been implicated in peptic ulcers and stomach cancer later in life. But it is also responsible for suppressing acid reflux, calming the inflammatory response of our immune system, and in regulating the levels of a stomach hormone that signals when to stop eating. With our traditional focus on "germs," the negative characteristics were discovered first, and the bacterium targeted for eradication. We are only belatedly learning what it does *for* us (when we still have it), and that the rest of Earth's inhabitants have it as a standard component of their makeup.

The Underground Economy

It should come as no surprise, at this point, that larger, multicelled organisms outsource much of their basic metabolic functions to the bacteria. They are the masters of metabolism. That is their specialty. The only thing surprising, until recently, is that humans aren't exempt from this arrangement. We are evolved to depend on them for our low-level interface to the biochemical environment. Like the registered guest workers and the undocumented aliens combined, they are the denizens of our underground economy that enables our more visible, official economy to function.

They do two things for us that we are not able to do for ourselves. First, because they have evolved specialists for breaking down and building up almost any possible molecular form that our energy and essential nutrients come in, we do not have to carry the huge load of genes that code for all of these catalysts in our own genome. We can use whichever of theirs come in handy in just the right environments. Second, we can exploit their much more dynamic and risky reproduction strategy to keep up with changes in our environment without risking our own conservative reproductive strategy. Recall that bacteria are very promiscuous, swapping genes wantonly and evolving at breakneck speed. Our genome cannot afford this because we have to preserve some very delicate relationships among trillions of cooperating cells, and tissue types. We can't evolve fast enough to keep up with all of the changes in our metabolic habitats. They can. And by using them, we benefit from their risk taking without being exposed to it ourselves. In addition to providing the invisible workers in our economy, they also supply the early stage entrepreneurs, taking on enormous risk, innovating at unsustainable rates, failing in huge numbers. We wait patiently for the winners to emerge, then make deals with them, sometimes just partnering at arms length, as in our microbiota, and sometimes acquiring them outright, as in our mitochondria. We just rent the 5 million plus genes in our microbiome, and avoid passing them through the internal inheritance channel of the 20,000 or so that we own. It is up to our partners to get the microbiome into our next generation.

Exactly what metabolic services our bacteria provide for us is still unclear. We are still trying to figure out which species are in which parts of the gut, and what their corresponding complements of digestive enzymes are. The general picture that emerges is that the closer to the stomach you get, the more our native genes are doing the work, and the further away, the more of theirs. It is known that our bacteria synthesize vitamin K, and some from the vitamin B complex – services which we cannot perform ourselves. But the main thing our bacteria can do that we cannot is to break down the complex carbohydrates that form plant cell walls. We can digest (and need) the simple sugars these are composed of, but without bacterial help, most of these will pass right through us, unused. The irony is that most of the bacteria that can digest these things are in the large intestine, past the point where we absorb most of our nutrients. So a lot of this fiber goes to feed them. But they provide other, very essential services for us down there, so it is important to feed them.

One of the problems created by the modern Western diet is that it has drastically changed the metabolic dynamics that we evolved under. We touched on this before in relating how our appetites are tuned to a much more impoverished rate of high-energy food availability. But even more critical is the *form* in which that food now comes to us. The large scale shift from whole foods to processed foods has made those simple sugars much more accessible to our native metabolism, since they are not locked up in the complex carbohydrates, and deprived our downstream bacteria of much of the food they evolved to expect. The nutrition industry still largely operates under the myth that the calorie content of a food is intrinsic to the food itself. It contains a certain amount of energy per gram weight. This doesn't take into account the amount of energy the human digestive system, our own enzymes and those of our microbiota, will *extract* from that food. Processing denatures proteins, homogenizes fiber, and breaks down complex carbohydrates. We can extract more energy from a gram of processed food than from the same gram of its whole food equivalent. Our original gut bacteria will get correspondingly less from the processed food and more from the whole food.

This creates an energy surplus for us that we are not evolved to deal with, leading to obesity. But it also changes the population of our gut microbes, giving us a mix that likes these simpler sugars and fats, and displacing the ones that thrive on fiber. As we learn more about this connection between our diet, our weight, and our bacterial population, we are beginning to discover some interesting feedback mechanisms. There is a reliable correlation between obesity and microbiome in humans. Lean individuals exhibit one mix of microbes and obese individuals exhibit a different mix. If an obese individual loses weight, the gut population gradually shifts back toward the other regime. But it is not just the difference in diet that drives this. When the gut microbes of obese mice are transplanted into bacteria-free mice, the recipients gain weight even when food consumption is reduced. These bacteria encourage the obese regime. This produces a latency effect between change of diet and change of gut population that may help explain why short term dieting often doesn't take hold and lead to a more long-lasting change in eating habits. It takes some time for your new diet to result in a change to your microbiota. In the meantime, the obesity-oriented species are still encouraging you to eat the old way.

These correlations also help explain why low-level antibiotic use, in animals raised for food, produces weight gain. This technique has been used for over 60 years by the food industry to produce larger animals, without any real understanding about why it works. Recent studies on mice show that antibiotics alter the gut population by killing off some native species, allowing other species to thrive. The changed population has an increased ability to extract calories from complex carbohydrates and produces more short-chain fatty acids. This is just one of the many worries that are emerging about the effects of antibiotics on our microbiome. Modern Western children receive an average of 10 to 20 courses of antibiotics by the time they are 18. Dentists prescribe them before oral procedures. In each case, we are targeting some particular species of pathogenic microbe, but we use a shotgun instead of a rifle. We take out the good with the bad.

There is more to the underground economy than just metabolism. In the lower intestine, where most of our bacteria reside and mostly feed themselves on the residue of what we don't eat, they feed us in a different sort of way. The epithelium, the layer of cells that line the inner surface of the lower intestine, serves as the semi-permeable boundary between the outside world and our inner metabolism. It serves to keep pathogenic microbes, their toxins, and other harmful things on the outside. But this tissue does not receive nourishment from the blood vessels like most of the rest of our tissue. Instead, it receives nourishment primarily in the form of the short-chain fatty acids that gut bacteria ferment from plant fiber. Without this maintenance, it can become too permeable, letting the bad things in that will cause an inflammatory immune response. There is even emerging evidence that our gut bacteria affect our brains and mental states. Changes to the gut population in mice have been shown to alter sociability and risk-taking, lower levels of anxiety, decrease stress hormones, and increase brain receptors. Who knew?

International Diplomacy

The traditionally hostile measures we have taken against bacteria are not entirely unjustified, of course. Some bacteria are not our friends. More of our human ancestors used to die in childbirth and of infectious diseases than today because of our recent sanitized environments, surgical interventions, and wonder drugs. But long before we learned about bacteria, and began to craft our hygienic and pharmaceutical defenses at the macro level, our vertebrate ancestors developed a much more fine-grained and nuanced defense posture in the form of our adaptive immune systems. Large, multicelled creatures can't do battle with prokaryotes using teeth and claws. There is a big scale mismatch. Bacteria are extremely small, they operate in hordes, and they evolve new defenses rapidly. So we developed a subsystem of our normally monolithic, tissue-based organization to deal with them on their level – rapidly evolving cell horde to rapidly evolving cell horde. This is our adaptive immune system, a huge army of fast-evolving, single-cell soldiers: sentries, reconnaissance specialists, and killers.

Since we ourselves are composed of trillions of cells, when we fight battles at the single-cell level, we have to be able to distinguish our own cells from those of foreigners. Our immune system does this by setting up perimeter defenses at the borders between our tissue and the outside world, in the form of sticky substances like mucus, and by posting single-celled sentries in our blood to recognize foreign cells that manage to get into our tissues. Once recognized, immune cells specific to the particular interloper are rapidly reproduced to mount a specialized army for cell-to-cell battle. If the invaders have already begun to multiply, this turns into an all-out, system-wide war. Inflammation and fever are triggered and the battles can rage internally for days before the enemy is subdued.

The adaptive immune system learns to distinguish self from non-self by producing huge amounts of immature T-cells that have a variety of receptors for recognizing all sorts of molecular side effects of cells, called *antigens*. These sentries are trained in the thymus (a small, two-lobed organ behind the breastbone) by being presented with many instances of antigens produced by the body's own cells. T-cells that recognize these self-factors are either eliminated or turned into *regulatory* T-cells that will later tamp down immune responses to these self-factors. The rest become *effector* T-cells that will trigger the immune response when they recognize their specific antigen. This basic scheme would work quite well in a world in which all non-self cells are the enemy, similar to the war-on-germs attitude that we adopted at the end of the 19th century. But the immune system evolved in the real world circumstances where vertebrates depend on communities of potentially friendly bacteria. It had to strike a balance. So its interface with the host's microbiota is very nuanced. There is a lot of diplomacy involved. Some of these guys are our friends, like the *lactobacilli*. Some are our enemies, like *Mycobacterium tuberculosis*. And some are at times friends and at times enemies, like Helicobacter pylori. Some are friends in certain places, but enemies in other places, like *E. coli*. Some are informants that carry on a constant crosstalk with the immune system across the borders. Some are double agents that disguise themselves with self-looking antigen coats. Many are neither intrinsically helpful nor harmful, but it is often good to let them colonize the DMZ to suppress other populations that are harmful.

This multifaceted immune system is our complex interface to our microbiota world. It matches them in size and variety and tries to eke out a mutual co-existence. Its posture is only occasionally one of all out war. It is more often one of containment and diplomacy, spy versus spy, trust but verify. We swap stories with them across the trenches. Bacteria aren't out to kill us. They are just trying to eke out a living in their environment, which just happens to be us. They mindlessly kill us from time to time, just as we mindlessly kill them (until germicides and antibiotics came along). The games are not always zero-sum. Sometimes we both do better when we cooperate. To appreciate just how exquisitely your immune system has evolved in maintaining this delicate balance, look at how quickly bodies decompose once they die. We call this decomposition, but it is really a big feast. All the little critters are eating you. Where did they all come from? They were always here – the barbarians at the gate, so to speak – waiting for an opportunity to sack the big food source. When death turns off the immune system, they come pouring in. Your continued existence as a living body is really a stable equilibrium between eating and being eaten, kept in balance by your immune system.

So what's a fledgling immune system to do when a new human is born? You are about to be massively invaded by thousands of foreign species. You can't start in a state of war because you will need some of these initial invaders. But which ones? The training of T-cells in the thymus marks them all for attack. Well, your developing immune system does something similar to what your developing brain does. Both know, through evolution, that they will be called upon to be sensitive to the contours of some aspect of the local environment. The immune system knows that this will be some local community of commensal foreigners, but not yet which ones. The brain knows that it will be the sounds of some local language, but not which one. So each goes through a tuning phase early on where it adapts to what it first finds. This early tuning will later be codified as the way to deal with that environmental aspect for the rest of the human's life. So the first microbial colonizers get a shot at being deemed the lifelong friends. Beneficial species that arrive too late, because the baby skipped the customary inoculation route, or got the wrong initial food, risk being classed as suspicious and provoking the immune response. We don't really understand how this works yet, but a recent study has confirmed for the first time that gut species can actually train the T-cells in the gut, not the thymus, by converting T-cells with receptors for their antigens into regulatory rather than effector T-cells.

***

This new perspective on life, of seeing yourself as a large pile of bacterial cells with a smaller amount of your own mixed in, is perhaps the mother of all gestalt shifts. It is jarring from a personal perspective because we've been conditioned to think of bacteria as creepy, crawly germs. They are pretty ugly and disgusting in the cartoons for personal hygiene products. But these are cartoons after all. Your bacteria are no more creepy and crawly than your own blood cells and immune cells that run around with them and interact with them all day long. You don't feel a thing. The only essential difference between "you" and "them" is in the DNA. You are a community, no matter how you look at it.

This is one of those shifts in the meaning of life that we are witnessing in real time. It is currently upending biological research, and just beginning to dawn on doctors in clinical practice. It is a conceptual shift that may have practical value for you in leading a better life, though the science is still too young to make specific prescriptions yet. This hasn't stopped the "probiotics" industry from rolling out products and claims, but there is almost no reliable science behind these things yet.

One lesson that this teaches is that we need to continually vet modern environmental interventions against the circumstances under which we evolved, to make sure we aren't breaking some essential ecology or other. It does not mean that all such interventions are hazardous. An across-the-board "back to nature" attitude will probably deprive us of more life improvements than avoiding broken ecologies. Our ancestors accepted hunger, pain, disease, and death at rates that would appall us today. What we need to do here, as with most issues involving cooperating life, is to learn how the ecologies work. They are complex things. With this new awareness of our bacterial ecology, we have learned that at least three things we used to consider inessential to modern life, vaginal birth, children playing in the dirt, and eating whole foods, are actually essential elements rather than dirty and inefficient vestiges of hunter-gatherer life that we can leave behind.

7 | Consciousness and Reality

True philosophy must start from the most immediate and comprehensive fact of consciousness: 'I am life that wants to live, in the midst of life that wants to live.'
— Albert Schweitzer

What's it like to be a bacterium? What would it feel like when your outer membrane is penetrated by a virus, it replicates itself wildly with your own reproductive enzymes, then blows you apart to release its progeny? Would it hurt? When the receptors in your outer membrane come into contact with glucose molecules, would you feel hunger? Would it taste sweet? Would you feel any affection for, or camaraderie toward, other bacteria of your kind as you are ineluctably pulled together with them in a colony? Probably not.

These questions ask us to imagine bacterial life from the point of view of human consciousness. Since it is impossible for there to be a human perspective at such tiny sizes, because humans themselves cannot possibly exist at that scale, there is no meaningful answer. Questions about conscious experience don't seem to make sense until we are large enough and complicated enough to have a nervous system. So even though an organism as primitive as a bacterium displays some of the outward behavioral signs that we normally associate with consciousness – moving away from toxins, moving toward food, associating with its comrades – we are not inclined to think that this goal-directed behavior is driven by an internal awareness of the external world.

But consciousness clearly does emerge from non-consciousness at a certain level of biological complexity. To see this, we need only contemplate our own embryogenesis. What's it like to be a zygote? What would it feel like to be a single-celled, newly fertilized human egg? This question makes just as much sense or non-sense as the question about a bacterium. Each of us was once one of these single-celled entities. If a bacterium doesn't have consciousness, neither did we. But in the course of nine months, each of us developed gradually, step by incremental step, into a multicelled creature with a nervous system that does have consciousness. Somewhere along the line, it emerged. It may well be that somewhere along the line even you, the person, emerged from a non-conscious, non-person. Were you really you when you were a zygote? We can ask the consciousness question at any number of stages on the way up. What's it like to be an embryo? What's it like to be a fetus? Somewhere along the line, the question makes sense. So either consciousness comes in degrees, or perhaps it's consciousness all the way down, but the lower you are on the scale, the fewer things you are conscious of.

We don't normally think of ourselves as having an inner awareness of the external world. We think we are *in* the external world. We are directly experiencing it. It is philosophers that remind us of the distinction between consciousness and reality. How do you know you are not dreaming? A modern version of this challenge asks how you know you are not just a brain in a vat – no bones, no tissues, no sense organs, just a central nervous system with your endpoint receptor and effector neurons wired to a powerful computer which feeds you a sensory version of an "outside world." Or, more recently, how do you know you are not in the Matrix? A curious thing about these thought experiments is that the outside world that your consciousness is disconnected from is always itself described from the point of view of human consciousness. It's always a human reality. It doesn't occur to us to wonder what it's like to be a dog. This is a sensible wonder. Certainly dogs are conscious. What's it like to be a horse, or any other herd/prey mammal with eyes on opposing sides of your head? Do you see separate versions of reality on each side with a gap in the middle, or does your brain somehow combine them? Since the two visual fields don't overlap, do you only see in two dimensions? What's it like to be a fly with multifaceted eyes? Or a bat with no vision at all, but echolocation radar instead that tells you essentially the same thing?

When you go up and down the animal kingdom, adding and subtracting senses, you come up with many versions of consciousness, and thus many versions of reality. This is a much more interesting philosophical perspective. Are there many incommensurable versions of reality? Or is there just one reality, and many conscious versions of it? In which case, what does the real one look like? We may be tempted to take an androcentric view of this, thinking that all of these other creatures' views are in some sense lesser, sensory-challenged versions of ours – that we see all of reality, and they just see incomplete or obscured parts of it. But we would be wrong. Other species have better visual acuity than us, see more colors, hear more sounds, smell more odors, and even sense things like electric and magnetic fields that we don't. Is our human reality just one among many?

Reconstructing Waves and Molecules

Well, one thing we have, that no other species has, is scientists – designated specialists who are able to think abstractly about the common causes of all of the experiences that any species has. Scientists are able to see themselves, as sentient beings, as just one of many elements of the environment whose conscious experiences are caused by other elements of the environment. From this third person perspective, you can see how your own conscious experience can at times be at odds with what is really out there, so you can't always trust it. But this perspective can also show you how to make your assessments of reality more reliable, and here is where humans pull away from the rest of the species. You form a theory about the causes of your conscious experiences, even causes that you can't perceive, that can reliably predict what the next set of experiences will be if you tweak the environment in some novel ways. Then you look to see if those next sets of experiences turn out the way you expected. This allows you to build up a more precise version of reality that you *can't* directly experience, that explains the things you *can* experience – a single version of reality that causes all the many different conscious perceptions of it, for all sentient beings.

This "least common denominator" reality is the ultimate gestalt shift, one that generally only particle physicists engage in. The current thinking is that this most basic reality consists of 40-something kinds of things, sometimes behaving like particles, sometimes behaving like waves. They are governed by four fundamental forces: the strong and weak nuclear forces that hold the constituent parts of atoms together, the electromagnetic force that governs most things at the molecular level, and gravity, which doesn't really come into play until you get to things with very large masses. Sentient beings and their conscious experiences don't emerge until you get to the molecular level, so the most basic biology of consciousness is a story of molecules and electromagnetic energy impinging on sense receptors. It is also the last version of the story that is the same for every species. The various *conscious* realities – the colors, the sounds, the smells, the tastes, the feels – are what we all make of these molecular goings on, even though none of us can directly experience them (except for humans with our artificial cyborg senses; see below).

When the ancient Greeks first proposed that the world ultimately decomposes into a set of smallest indivisible parts called atoms, they had in mind parts that were a lot like our familiar, observed objects, only smaller. One could imagine the color and feel of the larger things accruing to the ever-smaller divisions of each, right down to the final, indivisible sizes. Our modern atomic elements, on the other hand, are arranged into a table of 118 things that we have mostly never heard of (like beryllium), punctuated by a few things we do recognize, like carbon, gold, silver, copper, mercury. What makes gold gold at the atomic level is not its shinny yellow color and hardness, but the number of protons in its nucleus. Take one away and you have platinum instead. Add one and you get mercury. A gold atom is not itself gold in color (or any color, for that matter), but if you combine enough of them together into an ingot, they collectively are. If we start with one atom, and add one at a time, how does the "gold" property of our conscious experience suddenly emerge when we add the first atom that crosses the threshold? The answer, of course, is that the familiar gold properties, or the familiar mercury properties, are not inherent in the atoms themselves, but arise from the causal effects that the atoms have on the human nervous system. There are no such properties without reference to our potential experience of them. A single gold atom will reflect some photons of the appropriate wavelength, but the signal is way too small to trigger any response from the photoreceptors in our retinas. If there were a threshold quantity at which a collection of gold atoms first appears gold, that would be the threshold at which the collection first reflected enough photons to impress our nervous system.

Human consciousness is disproportionately focused on vision. We so favor this sense that we tend to describe direct experience in general as a form of sight, as in 'eye witness' or 'observation.' In physiological terms, this sense boils down to photon receptors in our retinas that are sensitive to a very small bandwidth of the electromagnetic spectrum: visible light. 'Visible' of course means 'visible to humans.' This concept is sometimes confusing because a number of dissimilar phenomena that we know by different names – radio waves, microwaves, infrared and ultraviolet radiation, x-rays and gamma rays – are all various bandwidths of this same spectrum. The spectrum represents a continuous scale of wavelengths at which electromagnetic energy can propagate. The longest wavelengths are radio waves, which run from a length of many kilometers down to about 10 centimeters. From 10 centimeters to 1000 microns, we call the energy microwaves. From there to a little under 1 micron, it is infrared. Visible (to humans) light in the familiar rainbow of colors from red to violet runs from about 0.7 microns to about 0.4 microns. From there to 0.005 microns is ultraviolet. Then x-rays to 0.00004 microns and gamma rays to 0.000001 microns.

With the huge variety of wavelengths available, you may wonder why we are sensitive only to this very narrow band of the spectrum, representing about 3 billionths of the total length. The answer is that energy in this particular band is extremely common in our Earthly environment. The ordinary objects that this energy enables us to track don't emit the radiation, they reflect it. So we need a constant source of external radiation to bounce off the objects and hit us in the eyes. We get this from proximity to our sun and our surrounding atmosphere. Our sun radiates electromagnetic energy mostly in our favored band, and our atmosphere edits out many of the wavelengths lying outside the range. So as a species, we have exploited the major range of electromagnetic energy that is available for reflective signaling. Since we get reflected sunlight for only about half of our 24-hour day, we are, in our natural state, a lot less effective at night, so we largely sleep it away. In our modern world, where we have manufactured our own sources of EM radiation to bounce off objects, like incandescent and fluorescent lighting, we are able to abuse our natural biology, and pay for it through sleep deprivation and jet lag.

Some of our compatriots in other species have made a successful living in the dark, during the absence of this electromagnetic irradiation, by exploiting the other major system we have available on our planet for long-range signal propagation: compression waves. EM waves can propagate in a vacuum, but compression waves require a physical medium: either air or water in the case of sound waves, or solids in the case of seismic waves. They are radiating compressions of the molecules that make up the medium. Ironically, we call the radio portion of the EM spectrum "airwaves" though they are nothing of the kind. Airwaves are sound waves. Human senses use air and water to passively receive sound waves originating from objects and events themselves, like voices, and sirens, and small collisions. Bats use sound like we use flashlights, generating the original waves themselves so that they bounce off ambient objects back to their receptors. Echolocation is a bat's "vision."

From the point of view of these two signal carrying systems, photons and ambient molecules, our familiar five senses really come down to two. We have one form of photon reception, vision, and four varieties of molecule reception, hearing, smell, taste and touch. In hearing and touch, what is important is the *pattern* of molecular collisions with our receptors. Our nervous systems don't really care which kinds of molecules are hitting us, just that some continuous stream of them are. In touch, it is patterns of *parallel* impact that are significant. Touch receptors operate as a group, similar to the photoreceptors in the retina, simultaneously reporting the differential intensity of impacts at millions of tiny points. The brain then computes an overall tactile experience of the kind of surface we have encountered. In sound, our receptors are sensitive to the continuous changes in the air or liquid wave impacting them, and the brain constructs an auditory experience by summing up the *series* of those changes. Taste and smell are sensitive to the identity of individual *kinds* of molecules, irrespective of the patterns in which they arrive. They are essentially the same sensory mechanism, a receptor that fires when it recognizes a molecule of a certain shape and charge, specialized to molecules in solids and liquids, in the case of taste, and airborne molecules, in the case of smell. At the molecular level, this recognition amounts to a kind of compatible geometric shape between the molecules of the receptor and the ambient molecule being detected. When the shapes and charges line up just right, a little bit of attraction occurs that sets the recognition events in motion. You are sensitive to as many tastes and smells as can be constructed out of combinations of the basic molecule recognition receptors you have in your nose and on your tongue.

The conscious experience available to us through this particular collection of five sense modalities, and the ranges of energy and molecules that they are sensitive to, is just one of the many conscious experiences available to inhabitants of this planet. There are many creatures with fewer modalities than us, such as earthworms, but there are also creatures with modalities we lack. The active echolocation sense of bats would be the equivalent of our having night vision by virtue of a spotlight embedded in our forehead. Bats forgo optical vision, but marine mammals like dolphins and porpoises, have vision *plus* the echolocation sense in their foreheads. Elephants enjoy a kind of "hearing" through their feet in which they detect the seismic waves of other elephants as much as 10 to 20 miles away, propagated through the ground. We also know that our sensitivity to sound waves is somewhere in the middle of the range of known possibilities. The big mammals, like elephants and whales hear at frequencies below our range and dogs can hear at frequencies above us. Various species of aquatic animals can detect changes in electric fields, either passively or by generating the fields themselves. Other species can sense magnetic fields. We can imagine what the consciousness of other species that lack some of our sense modalities must be like through simulation, by covering our eyes and ears. But what would the integration of echolocation and vision be like? And what on Earth would the features of electric or magnetic fields feel like?

Experiencing the world by recognizing its individual molecule types, what we think of as taste and smell, is the most ubiquitous way in which organisms of all shapes and sizes get around in it. It is the primary sense modality of the truly tiny creatures like bacteria and fungi, and it continues right up through the various cellular parts of ourselves. It's how the immune system recognizes pathogens, how the endocrine system regulates the body with hormones, how digestive bacteria in our gut recognize their specific foods, how the gene regulatory system turns its parts off and on with proteins. Because we have taste and smell, we have some idea of what this view of the world is like, though it is hard for us to appreciate what it must be like with only this kind of experience. Of all the many different ways of experiencing the world that Earthly species have come up with, we can directly relate to only a few.

Consciousness Evolves

Unless we are willing to grant consciousness to the sensory reactions of single cells, consciousness is a feature of nervous systems. Consciousness is an interpretation of the outside world that brains put on the signals coming through sense receptors. So the simple peripheral nervous systems of creatures without brains, where the sensors are wired pretty directly to the effectors, probably don't support consciousness either. Vertebrates, the family of animals whose nervous system design we share, don't start until there is already a semblance of a brain and a central nervous system, so they form the baseline for our own version of consciousness. Our brains and sense organs still carry vestiges of most of the parts of our vertebrate ancestors', and these common parts suggest that the seat of consciousness has moved as brains have evolved.

In creatures more primitive than vertebrates, like jellyfish, the wiring between body sensors and body effectors is pretty direct. There isn't a lot of neural processing in between. The basic vertebrate nervous system rewired these direct connections up through the spinal cord to the brain where it meets the wiring from the long-range senses. This gives the brain a chance to override the contact sensors and effectors on the surface of the animal based on input about things still at a distance. You don't have to wait for (large) things to hit you. You have a "picture" of the outside world telling you where all of these eventual things are located. The evolution of our three long-range senses that provide the input for that picture – smell, sight, and hearing – were the likely cause of the emergence of the three basic brain segments in early vertebrates: forebrain, midbrain, and hindbrain. Anatomically, noses precede eyes, and eyes precede ears in animals' heads, and sensors from noses are wired into the forebrain, eyes into the midbrain, and ears into the hindbrain. As animals' brains have become more complicated, these brain regions have taken on additional functions, and the processing of the three senses has been spread out into multiple regions by additional interconnections, but at the lower end of the animal complexity spectrum, the three brain segments are nicely paired with the three senses. We have a lot more brain parts than the original three sections, but vestiges of those sections still survive in the now diminutive central stem portion of our brain. Our three long-range senses are still wired through these parts, and our sense organs still operate by the original principles, though the outward anatomy of our ears bears no resemblance to fish "ears."

What Ears Detect

Our human sense of hearing, as an interpretation of compression waves in the air, as well as our sense of balance, acceleration, gravity, and orientation in space, which are also detected in the inner ear, evolved from surface sensors called *neuromasts* on our aquatic ancestors. Most fish have two lateral lines of these hairy cells running along each side of the body. The movement of the hairs against the water detects flow, pressure and vibration. These sensors are wired directly into the hindbrain where the animal computes its orientation in the water, and motion in the water, as well as the ambient motions of nearby creatures. Over evolutionary time, these lines of neuromasts have been internalized inside the animal where they continue to sense the movements of fluid, only now the fluid is inside its own semicircular canals – our inner ears. The lagging motion of these now internal fluids as we move, rotate, and accelerate our bodies in space, moves the now internal hairs lining the canals and sends similar signals to what is now part of our midbrain. There we interpret it as balance, movement and spatial orientation.

We interpret the pressure patterns of airwaves that cause those hairs to move, as sound. These signals are further routed up into our primary auditory cortex where we process the features of sounds that are significant to humans. Language sounds are further interpreted primarily in the left hemisphere and music sounds primarily in the right hemisphere. Sounds, particularly the culturally meaningful ones like language and music, are the primary focus of our auditory consciousness. We tend to forget about the companion balance and orientation-in-space experiences that our ears and midbrain provide. When we are paying attention, these are a kind of conscious experience, but not one we associate with any of our "five" senses. The circuits connected to these sensors in the midbrain are capable of orchestrating appropriate responses to the "movement in space" data without our conscious intervention, such as automatically sticking out our arms to break a fall when we lose our balance, or automatically turning our eyeballs in the opposite direction as we turn our head so that we can maintain our gaze in a fixed direction. In the falling reaction, your brain is intervening in the default spinal cord movement program, but from your conscious point of view, it might just as well have been done in the spinal cord because you weren't consulted.

These sub-sound experiences are not (normally) part of our conscious experience. An interesting question is: are they just *no longer* part of our conscious experience? Are they perhaps still part of the consciousness of fish that don't have our higher brain regions? We can only speculate about what a fish's unified sense of neuromast movement is like, without the sound part being peeled off and routed to a higher processing center. Has our consciousness migrated up to the cerebral cortex so that we are no longer conscious of midbrain experience, or do fish simply have a more impoverished sort of consciousness of sounds, similar to our vague sense of location in space?

What Eyes Detect

Our human color vision is *trichromatic*. Our brains construct the endless varieties of color that we experience by sampling our small slice of the EM spectrum at three of the wavelengths: between violet and blue, around green, and around yellow. By comparing the firing rates of receptors for each of these wavelengths at each point on the retina, the brain computes our variety of color experience at that point. You can see an artificial example of this process at work by looking at a video display under magnification. You would see, at each pixel point, a trio of tiny lighting elements, one each for blue, green and red light. The TV signal, or the video bitmap on your computer display, has already encoded the scene into a particular intensity (or absence) of each of the three colors at each pixel. The overall image that you see without magnification is an abstraction of all these little pixels by your brain into the continuous variety of colors approximating those of the real world. The same principle, with a different trio of primary colors, is at work in our perception of printed color. If you magnify a color photograph from a newspaper or a printed brochure, you will see lots of tiny dots of cyan, magenta, or yellow (and sometimes black). When you back away, you see the full spectrum of colors. Exploiting your trichromatic nervous system like this enables the publisher to print the full spectrum of colors without having to use the full spectrum of inks. Three will do. You get a clue about this phenomenon when some pages of your newspaper are misaligned during printing, causing the three colors to be slightly offset. You get a blurred picture with cyan, magenta, and yellow ghosts around the edges. The individual dots in color printing, and the pixels in color video, are gargantuan compared to the size of your retinal photoreceptors, but they are both sufficiently below your brain's resolution threshold to give you the perception of continuous color variety.

We are one of the better discriminators of electromagnetic radiation among the mammals, most of which, other than old world primates and some new world primates, have *dichromatic* vision – photon receptors for only two different wavelengths (similar to a color blind human). But we are by no means the best. Turtles, some fish, and most birds have *tetrachromatic* vision (discriminating four separate wavelengths). This allows them to discriminate orders of magnitude more colors than we. There are even species, pigeons and butterflies for instance, that possess five or more kinds of photoreceptors in their retinas, making them potential *pentachromats*. Primates, like us, are descended from non-primate mammals, which are in turn descended from fish. A curious thing about the evolution of color vision in this lineage is that tetrachromacy was the original standard. Reptiles and birds kept this and improved on it somewhat. But on the mammalian branch, the variety of visual reality has receded. Early mammals lost two of the four pigments of their predecessors, then we got one of them back. Mammals first emerged as small, nocturnal creatures in the age of dinosaurs, so their echo-niche did not require great color vision, leading to its atrophy. It is thought that primates evolved one of the pigment receptors back because of our arboreal, fruit-eating lifestyle in which it is useful to be able to distinguish red fruit from green leaves.

Just as our sound perception is in the middle of the perceived compression wave spectrum, our vision is in the middle of the total perceived range of electromagnetic wavelengths. Pit vipers can sense in the infrared range, and bees and birds can pick up the ultraviolet. We can approximate what things look like in the infrared and ultraviolet ranges by recording photos that pick up these wavelengths, then mapping them onto colors we *can* see – using red for infrared and violet for ultra violet. This allows us to see otherwise invisible patterns on flowers that bees use as pollen beacons and elaborate markings on birds that they use in mate selection. But by doing this, we are cheating. We won't be able to see any ultraviolet/violet distinctions in these reconstructed photos. We can get a feel for what the dichromatic experience of other mammals must be like by artificially reducing the color variety in a picture. Imagine trying to explain to a color blind human what red and green are like. What could you possibly say? You just have to be able to experience them. Now imagine being on the other side of the disparity. Even if birds could talk, what could they possibly tell us about some of their colors that just don't exist for us?

The ancient wiring from your eyes to your midbrain has been augmented and superseded by wiring that goes on to your visual cortex at the back of your large cerebral hemispheres, but a lot of the visual processing still goes on in the midbrain part of the brain stem. And like other stem functions, it happens mostly below the level of conscious awareness in humans. When objects suddenly appear in your visual field, or move precipitously toward your head, you automatically execute the flinch reaction, whether you want to or not. When provocative objects appear in your peripheral vision, you sometimes involuntarily turn your head in their direction. In these cases, your midbrain is exercising its override prerogative over the spinal cord, on behalf of the whole brain, by executing a hardwired response programmed eons ago by evolution. Your brain was in control, but not the part of your brain that "you" control.

But at least you are consciously aware of the visual stimuli that cause these involuntary reactions, if a bit late. This is because the visual input also reaches your higher, visual cortex, even though the midbrain acts to produce the response before your cerebrum gets a chance to weigh in. In people who are effectively blind because of damage to the visual cortex, or to the information pathways feeding into it – people who report having no conscious experience of sight at all – but still have functioning eyes and midbrain, a phenomenon known as blindsight is often reported. When presented with lights or colors or shapes in various positions in their visual field and asked to identify them, they report seeing nothing at all. When asked to just randomly grasp at where they imagine the light might be, or randomly guess the shapes or colors, they are successful much more often than chance. Some are accurate in almost every trial. One such person was asked to walk down a hallway that, unbeknownst to him, was littered with randomly placed obstacles. He deftly navigated around the obstacles without any conscious awareness that they were there. The intact midbrain is still receiving the visual input and orchestrating appropriate bodily responses, even though the visual cortex is now out of the loop. This is the same part of the brain that a frog uses to track and catch flies with its tongue. We can't say that blindsight patients do not really "see" objects without also saying that frogs don't really "see" flies. All we can say is that blindsighters are not conscious of seeing the objects they see. We have no reason to say the frog is not conscious of seeing the flies. Whatever consciousness is, perhaps it was once seated further down in the brain stem when that was the dominant part of the brain. Perhaps consciousness moved farther forward as higher regions evolved to exert control over lower regions.

What Noses Detect

The olfactory bulb is the brain's dedicated computer for analyzing smells. Receptor neurons in the nose run their axons directly into the olfactory bulb. These neurons differ from those used by the other senses in that they die after a few months and have to be constantly regenerated, including their axonal wiring back into the olfactory bulb. Each such receptor has a particular recognition molecule that binds to a particular kind of small molecule in the substances that we smell. About 3 percent of the protein-coding genes in mammals (including us) are devoted to making these smell receptors, though many of ours are no longer functional (see below). The 5 million receptor neurons in our noses are composed from 387 different kinds of recognition molecules. That means we can detect 387 different primitive odorant molecules, but these are not the things that we actually sense. The ambient molecules that we typically smell are large and complex with many sub-regions. Our relatively small set of primitive receptors bind to these smaller molecular regions, so a given substance will typically set off a number of these primitive recognizers. All receptors for the same type are wired to a common location in the olfactory bulb, and these primitive recognition centers are then interwired for pattern analysis. So what gets reported out of the olfactory bulb into the other parts of the brain that use it are these combinations. We smell the patterns. This gives us a repertoire of final scents far in excess of the 387 front-line recognizers. Recent research estimates that we can distinguish at least a trillion of these different patterns. Think of the primitive receptors as recognizers for individual letters of the alphabet. The olfactory bulb then computes spelling, and what we ultimately smell are words.

The olfactory bulb is at the front of the original vertebrate forebrain, so the direct wiring of noses to their forebrain is still with us, although in humans the role of olfaction has become so diminished relative to everything else that gets computed in our forebrain that this connection does not stand out. Not so with most non-primate vertebrates. When you look at the brains of our vertebrate cousins, from the most primitive hagfish and lampreys, through the cartilaginous fish, the ray-finned fish, the lobe-finned fish, the lungfish, the amphibians and many mammals, you see a large and prominent olfactory bulb at the very front of the brain that is often as big as the rest of the endbrain itself. The smell sense was a very big deal for a very long time. Humans still have olfactory bulbs but they are tiny and buried deep below the cerebral cortex.

The original nocturnal mammals had much more use for the sense of smell than we have today. Although we inherit many of their genes for producing types of odor receptors, about 60% of ours are no longer functional. It is thought that primates' increasing reliance on sight over smell has relaxed the selective pressure to keep these genes from being broken by mutation. It is also thought that given the rate at which we have lost them, humans are likely to continue losing them in future evolution.

In the sense modalities, like olfaction, that detect the presence of molecule types, as opposed to the patterns of their transmitted energy, there is no notion of spectra for interspecies comparison, but there is an analogy with color vision where the number of different colors you can discriminate is a function of how many different wavelengths your receptor types can detect, and how many receptors of each kind are bunched together on your sensory surface. Our 5 million odor receptors from about 387 types compares with mice, which have about a billion receptors from 1035 types, dogs (in general) with about a billion receptors for 811 types, and bloodhounds in particular, with about 4 billion receptors. For us, olfaction is not really much of a long-range sense anymore. We might use it occasionally, for example, to locate the source of something that is burning, but we generally don't follow scents. We use them to detect the near-range features of things such as food, as well as to augment our contact sense of taste. For bloodhounds, though, it is very much a long-range sense. A trained bloodhound can continue to detect the olfactory trail of a particular human that is 2-3 days old.

Partly because the olfactory sense is so ancient and has its own dedicated, hardwired recognition system, we are not very articulate about smells as a conscious sense. We are aware of them, but their processing is largely subconscious. We know that the smell sense plays a more prominent role in the lives of lower animals than in humans, particularly regarding the recognition and avoidance of sexual partners. Many animals have a specific vomeronasal organ in their nose that orients them toward potential mates during the appropriate periods in their reproductive cycles, and helps them avoid mates that are too genetically similar – an incest avoidance mechanism. It is sometimes claimed that humans still retain vestiges of this system, and some experiments seem to confirm that human females have a heightened sense of smell during ovulation, and have preferences for opposite-gendered partners that align with differences in their subliminal scents. If true, this would be the olfactory equivalent of blindsight for us – our brains sensing and acting on our behalf, even though we are no longer consciously aware of the sense itself.

Guessing at Reality

From the third-person perspective, where you can see both the observer and the thing observed, say at a magic show where you are in a position to see the sleight of hand that the audience cannot, you are aware that visual consciousness can be at odds with what is actually occurring in the world. But we tend to think that this kind of illusion is rare – that visual consciousness is largely a direct reflection of reality. What the audience actually witnesses is the successive movements of hands and things, which actually occur, and then they *infer* a cause that they don't actually see: magic. Little do most of us realize that this sort of informed guessing constitutes most of what we think we see directly. Your nervous system is pretty good at this. It takes some samples of what's out there, compares these with what has typically accompanied such samples in the past, and infers the rest. You don't see the evidence and then form the conclusion; you actually see the conclusion.

There is no such thing as the immediacy of perception with the long-range senses. They are all based on signal propagation from objects to us. Signals take time to travel. We think vision is immediate because the speed of light is so fast at close range that we can't detect it as a speed at all. It's effectively instantaneous. But when we look at the night sky, this speed is much more relevant. What we think of as present stars are actually the light from distant stars that may have disappeared millions or billions of years ago. They are not really there. It takes light a little over 8 minutes to travel from the sun to us, so the sun we "see" may actually have exploded 7 minutes ago.

Even here on Earth, though, where the speed of the signal is not significant, there is still a bit of a bandwidth problem. The photoreceptors on our retinal surface are not uniformly distributed. There is a higher concentration, allowing for greater acuity, in a small area in the center, called the *fovea*. This is the portion that can perceive fine detail. Only a small part of the total visual scene falls within this focus. To get good resolution over the entire scene, the brain moves the eyes in a series of rapid shifts, called *saccades*, to focus the fovea on different parts of the scene. These movements are faster than you can perceive, so you are generally not aware of them, but they still take time. Your brain presents you with a stable, high-resolution image of the whole scene by continuing to "broadcast" older versions of the parts of the scene that have not yet had a saccade update. It is really guessing that nothing important has changed between saccade intervals. So your "immediate" vision is really a kind of moving average of things occurring over an interval of time. Your brain is sampling the visual field at particular points and constructing the "current" scene from the current sample plus the memory of recent samples. You can experience this illusion by looking at your own face in a mirror and switching your gaze back and forth to look at one eye or the other. You will see no eye movement. Someone else watching you will.

The targets of the saccades are generally areas of the scene that are of most current interest to you – the ones you want real time updates for. A more generalized version of this selective seeing occurs when your brain simply edits out whole portions of the scene from your conscious perception in order to concentrate on some areas of interest. This illusion can be seen in experiments where subjects – whole rooms of them even – are shown a video of two teams of people, wearing two different colors of shirts, moving around a room passing large balls back and forth to teammates. The subjects are asked to count the number of passes between one of the two teams. Subjects who pay enough attention to correctly count the passes do not see a provocative character that walks right through the middle of the scene during the passes. This is sometimes a person in a gorilla suit, and sometimes a girl carrying a large umbrella.

Our brains have evolved to construct a version of reality that is most conducive to our survival. So even within the narrow bounds of energies that a given species interprets as its reality, each individual member of that species is conscious of a further custom version of that reality tailored to its own particular needs.

Cyborg Senses

One of the benefits of being the species that figured out the common reality behind all of these many conscious versions is that we are in a position to do something about the parochialism of our native senses. We originally had to theorize our way to the things we can't see, like cells, and molecules, and atoms. Now we can turn around and use those theories to build instruments that let us actually see them. And 'see' is the operative word here. Given the choice, we always produce a *visual* representation of the unseeable realities, even when optics doesn't apply. And why not? This is our favorite sense, and we can reinterpret the data anyway we please. We have now taken evolution into our own hands, manufacturing artificial improvements to our visual consciousness to make up for what we lost to the birds and the reptiles. We can't regain the missing colors, but we can use the ones we have to color aspects of reality that only we can see.

We haven't yet got cyborg senses of the kind that are surgically embedded in our bodies, but this is because we don't really need to embed them. We can just wear them during the times we need them.

And we typically don't need to wear them; we can just walk up and look through their viewfinders, or look at their screens, or have an image sent to us via email. All of these external appliances become extensions of our biological senses.

Our visual augmentation technology began, ironically, with the very pragmatic need to bring the deficient eyesight of many humans up to the normal functioning human standard. This culminated in the invention of glass lenses for eyeglasses in Italy toward the end of the 13th century. By the 14th century, the Italians had perfected the art of lens grinding to more precisely control the shape of lenses. This led to a generalization of the lens as a device not just for correcting vision, but for indefinitely enhancing it. Toward the end of the 16th century, the Dutch put this lens grinding technology to use in both the first telescopes and microscopes. The range of new visual experiences that these devices enabled, at both very distant scales and very small scales, eventually led the scientific elite to realize just how parochial our perception of the universe had been. Discoveries of the really big things came first, changing our view of our place in the cosmology. Discovery of the really small things came later, first with the curious microscopic properties of ordinary inert things, like cork, but eventually with the truly revolutionary discovery that a whole new order of living things had been earning an undetected living right under our noses (and in them as well).

Both of these technologies, telescopes and microscopes, were conceived as ways to increase our resolution of the portion of the electromagnetic spectrum, visible light, that we were already equipped by evolution to be aware of. We essentially copied the technology of our own natural lenses to better focus a familiar spectrum of photons on our retinas. We improved our ability to analyze the data we were already receiving, rather than expanding the range of electromagnetic energy we could sense. This trajectory, of course, had a lot to do with our initial conception of light as its own natural kind, as opposed to our current view of it being but a middling slice of a much larger continuum. Both of these scopes eventually reached practical limits for the magnification of visible light. At the telescopic end, these indeed were just practical limits. We were limited by the sheer size of the lenses and mirrors that would have to be manufactured for a stationary position on some surface of the Earth. We could, of course, always improve the resolution by putting those same size lenses out in space to increase our proximity to the distant objects we are trying to magnify, and eliminate all sources of interfering ambient light. So eventually we did. At the small end, though, we eventually ran up against the resolution boundary of the wavelength of visible light itself.

By the time we realized that visible light was just a piece of a larger spectrum, and thus that we could increase our sense of the spectrum at both the small and large end with other kinds of reception gadgets, we were already thoroughly immersed in the habit of calling our instruments telescopes and microscopes, so we just kept on doing that. We now have electron-tunneling microscopes that enable us to "see" in spaces too small for optical lenses. We have radio and microwave telescopes that enable us to "see" the longer end of the EM spectrum. We have subtly switched from the artificially restricted technology of improving our lenses to inventing whole new forms of vision in which some device either creates measurable disturbances in the really small things we want to observe, by some known principles of physics, or detects the long EM wavelengths radiated from the really large, far off things we want to observe, and another part of the device, usually an embedded computer, reconstructs a visual representation of these extreme things at a resolution we can see with our normal eyes. Our instruments now mimic the whole cycle of natural perception from the reflection or radiation of photons at the target, to their capture on the retina, to their processing and ultimate integration in the brain. We still call some of these devices microscopes and telescopes, where their use is similar to the old optical variety, but the more accurate description of our new plethora of scopes is "artificial receptors". We have targeted all the known ranges of electromagnetic, seismic, and molecular energies and built both new receptors for sensing them, and new computational paradigms for analyzing and presenting it to us. Not surprisingly, almost all of our artificial presentation media are visual. We have created whole new ways to "see" any kind of energy that can be detected. We have also improved on the sonar-style sense of some of our Earthly compatriots by generating our own energies to send into and through objects, sometimes measuring the reflections that come back, and sometimes measuring the differential amounts that reach the other side (x-ray films for instance).

By continually reinventing these new fangled ways of seeing, we are now able to see hitherto unseeable things such as the elementary particles emitted from high-energy collisions with larger particles. We can see super nova explosions that occurred millions of years ago on the other side of our galaxy. We can see the patterns of cosmic background radiation that occurred near the origin of the universe. We can see heat-producing animals in complete darkness by sensing their infrared radiation. We can see the invisible ultraviolet color patterns in nature that birds and bees see. We can see the geographic layout deep inside the Earth's crust and at the bottom of its oceans. We can see deep inside human bodies, see the structure of their soft tissue, see the real time activity in their brains. We can see the patterns of gene expression in developing embryos, and the activation of specific neurons in the brain, in phosphorescent color. We can see the real time activity of the nanoscale molecular machines that carry out all of the business of life.

This puts a whole new perspective on conceptual musings about "what X would look like," for all of the Xs that are too small, or too far away, or out of our energy band, or hitherto only described by mathematical models. In one sense, many of these things wouldn't look like anything, because there is no physical scenario under which we could possibly look at them directly. But since our native seeing is really a conscious reconstruction of things anyway, these new technologies are just another, more expansive way of reconstructing visual images. How much leeway do we have when we construct a visual image of unseeable things, you may wonder? Well, there are essentially two aspects of visual presentation: shape and color. Most of these unseeable things were previously just mathematical models. Mathematics doesn't have color, but it does have shape. So the geometry of the mathematics constrains shape pretty well. Although atoms are not solid, the tiny universe of their parts carves out a spherical shape (one of nature's favorite formations). So representing them as spheres is pretty true to reality. Coloring them blue, on the other hand, is arbitrary. The same for molecules. They are still too small to have meaningful colors, but we can model their elaborate shapes fairly accurately. We have taken to using colors to illustrate some of their properties, like charge and protein domains. The choice of colors is arbitrary, but if consistently applied, it is a very accurate way of seeing the different values of those properties. It's the same thing we have been doing with geographic maps and globes for centuries. We regularly use "heat maps" – continuous variations in color from blue to red denoting less of more of a property (sometimes heat!) – to illustrate the spatial distribution of variations in a property.

Choosing visible colors to represent the "invisible" portions of the EM spectrum is also somewhat arbitrary, but a little closer to actual perceptual consciousness. Some species see some of these variations in wavelength as variations in colors. We don't have enough colors, so we have to reuse some of ours. This creates some interesting side effects when we represent the images from space telescopes. Prior to the Hubble space telescope, humans were accustomed to two sorts of published images of outer space: a lot of blackness punctuated by the small light points of stars (pretty boring), and artists conceptions of close-up looks of stellar phenomena in dramatic colors (wow). When some of the first Hubble images of gas clouds and star nebulas were published, a lot of us assumed we were looking at more artists' conceptions. They certainly were beautiful, inspiring scenes, rivaling creation dramas of Renaissance artists. Upon learning that, no, these are actual telescopic images of *real space*, we were stunned! Wow, space looks like that?

Well, it might if we were able to see in the infrared and ultraviolet parts of the spectrum, and could see invisible variations in the single colors from the visible part of the spectrum, not unlike what some other species can see. Hubble captures the images in black and white, and NASA chooses the color filters to reprocess the picture for us. They use the standard red, green, and blue filters of trichromatic color to enhance what would otherwise be too faint or invisible contrasts for us. Some of what we are seeing would actually look that way if we had this kind of super resolution. Some of it is artificial, but still constrained by actual variations in the spectrum. Infrared is the reddest red, for instance, and ultraviolet is the bluest blue. Distinctions in visible light that we just don't have enough native colors to discriminate are enhanced by putting variations from the same part of the spectrum through the three separate filters. The light coming from oxygen, hydrogen and nitrogen atoms, for instance, all falls in the red spectrum, so we can't see the differences. NASA does us the favor of filtering each of the three light sources through one of the separate RGB filters. The shapes are real.

***

This meaning of life is one of those impractical ones that doesn't have much consequence for our daily lives. What's relatively new from biology is our enhanced understanding of how our brains actually form our conscious images, and the very different perspective that other Earthly species must have on this process, given the differences in their brains and sense receptors. This has the potential to upend some traditional philosophical views about consciousness and reality, but for normal folks, it is not a very pressing issue. Knowing what reality might be like from the point of view of a non-human consciousness won't profoundly change your life, or lead to a cure for cancer, or help you find your car keys. But you may just idly wonder about it. It puts your own consciousness of reality into perspective. Here's a thought experiment. To imagine what consciousness must be like if you interact with the world only by touching it, like an earthworm say, try to imagine what Helen Keller's consciousness must have been like. She was blind and deaf from birth, so she had no reference sights or sounds to remember. Since our sense of smell is not much of a long-range sense anymore, she was pretty much restricted to the contact senses – what she could taste and feel. Yet she managed to learn language and to carry on a very productive public life in a world where people and things are built out of smells and tastes and touches. To imagine what a bird's richer vision might be like, relative to ours, think of the difference between the Hubble, spectrum-enhanced photos, and our normal view of the night sky.

8 | A Diversion

If a tree falls in the forest

If a tree falls in the forest, and no one is there to hear it, does it make a sound?

Yes.

There really is no anomaly in saying that real objects have real macroscopic properties, even though no one is around to experience them, for once we know what it is about such objects that causes us to experience them the way we do, we can always identify the macroscopic properties with the causes. The causes are always there, even if some of the effects are not. So when a tree falls in the forest, and nobody is around to hear it, it does indeed make a sound. The percussion of the fall propagates a compression wave through both the air and the ground. The wave is there, even if no humans are. This well-known scenario is typical of "paradoxes" that are only paradoxical if you assume that only humans are conscious. It is extremely unlikely that the sound is not perceived by *some* sentient creature. Various mammals, like squirrels, and any birds will hear the airborne wave. Earthworms will perceive the seismic wave propagating through the ground.

To complete the story, when an external door on the international space station suddenly slams shut, and no humans are around to hear it, it doesn't make a sound, because there is no atmosphere to transmit a compression wave. So it also doesn't make a sound *even if* several humans, tethered on a spacewalk, are there to hear it.

Next question.

9 | Running Life Backward

Ah, but I was so much older then. I'm younger than that now.
— Bob Dylan, *My Back Pages*

Time runs in only one direction – forward. Time travel into the future is feasible because of relativity, but time travel into the past is not. Traveling backward in time doesn't quite contradict any physical laws, if your trajectory is self-consistent (you don't kill one of your ancestors, for instance), but we have no idea how this could possibly happen. Biological time is also a one-way process. Things evolve and grow and mature; they never unevolve or ungrow or unmature. This doesn't stop us, of course, from imagining stories in which people travel backward in time, or in the biological case, in which people grow younger. In *The Curious Case of Benjamin Button*, the main character is born a 70-year-old, then spends the rest of his life getting younger. It is not particularly feasible to be born an adult, because you have to be infant-sized to fit in the birth canal, but then inexplicably become very big, very fast. A more feasible scenario would have a normal infant growing to an adult and then turning around and growing younger again.

This would be physically possible, but not very likely. A human grows from a single cell, by cell division and differentiation. To reverse the process, your 37 trillion cells would have to successively re-merge with each other and change back into their original undifferentiated states, backing you up to a smaller and smaller body. The merger of two cells makes a bigger cell, so after each merger, the combined cell would have to shed resources and get smaller – not impossible. It just doesn't happen that way in nature. Even leaving out the part about shedding resources, cells just don't transform from a more differentiated state, like muscles or neurons, to a less differentiated state, like an embryo.

Until now.

From Out of Nowhere

It takes an average of almost 18 years for a scientist to win the Nobel Prize for physiology or medicine after the initial groundbreaking discovery. This is because groundbreaking discoveries are often so amazing that they turn out to be too amazing. They can't be reliably reproduced by later scientists. In cases of groundbreaking theories, it takes a while for experimental confirmations to roll in and for the scientific community to reach a lasting consensus. In the case of Shinya Yamanaka, it took only 6 years. His discovery of *induced pluripotent stem cells*, iPS cells for short, in 2006 got him a Nobel Prize by 2012. It could have happened sooner. In the intervening 6 years, there was very little doubt in the scientific community that this discovery had fundamentally altered our understanding of cell and developmental biology. What Yamanaka discovered was that mature, fully differentiated adult cells, such as skin cells, or muscle cells, or neurons – any cell, actually – can be driven backward in biological time to an embryonic state from which it is capable of becoming any other type of cell in the animal's body, or a whole new version of the animal itself. Replications quickly followed, better techniques were quickly discovered, whole new areas of medicine and research were spawned, and the first clinical uses in human trials are about to begin. This was disruptive (in the good sense).

To understand how disruptive, you have to appreciate that for biologists this was kind of like the discovery of alchemy – as a *science*. From our present perspective, we think of alchemy as a medieval pseudo-science, a mixture of superstition and wishful thinking – searching for the magic procedure to turn things into gold. But placed in its original context, alchemy was just an immature precursor of chemistry. Isaac Newton was a practicing alchemist. In the modern theory of chemical elements, the difference between elements is a difference between the number of protons in their atomic nuclei. Gold has one more than platinum and one less than mercury. So you actually could turn either platinum or mercury into gold by adding or subtracting a proton. The problem is that it takes infeasible amounts of energy to do this in a lab. So platinum and mercury stay put. But we know that nature does this on its own in high-energy reactions in space like supernova explosions. So the transmutation of elements isn't science fiction, it just doesn't happen on Earth.

The transmutation of cell types happens on Earth only in one direction. And we still understand considerably less about what makes two cell types different than what makes two atomic types different, even though the unidirectional transformation of cells happens right in front of us. It was certainly reasonable to assume that since it *never* goes backward in nature, there must be some fundamental biological reason for this. So Yamanaka's discovery was a sort of alchemy. He didn't discover nature doing this, something everyone else had just missed. He discovered the magic formula for intervening in nature to *make* it do this. To make it violate its own immutable laws – and it still works! The forward direction of biological time was apparently neither immutable nor a law, just one of the ways development *can* work. It can also work in other, radically different ways, even though it never has (that we know of). Nature has stunned us before with its robust modularity (DNA, the Hox genes), but this is crazy. And like true alchemists, researchers were quick to use the magical formula to make medical gold.

Yamanaka's lab started with 24 genes known to be active and important in the functioning of embryonic stem cells. These are the cells that form soon after an egg is fertilized, and are capable of becoming any of the eventual cell types of an adult (this is the *pluripotent* part). Many of these embryonic genes will have been turned off, or tamped down, by the time these cells differentiate into mature adult tissue-type cells. So the aim was to see what would happen if you could somehow get these genes turned back on in a mature cell. We don't know how to undo the switches, so the researchers tried the next best thing: put new copies of these genes, that didn't go through the turning-off procedure of prior development, into the mature cell artificially. They began with mature mouse fibroblast cells (cells that make the connective tissue that holds other animal cells together) and inserted the 24 new genes into the cells' DNA using retroviruses (this is the *induced* part). (This is how viruses normally work – by integrating their own DNA into yours). Sure enough, some of the resultant cells expressed some of the characteristic marks of embryonic stem cells. So they repeated the process many times, eliminating one or more of the original 24 genes to find the smallest subset that was both necessary and sufficient to produce the embryonic properties. They found that 4 would do it. To test for pluripotency, they implanted the reprogrammed cells into female mice but could not get them to grow full-term to viable offspring.

That was in 2006. A year later, three independent groups demonstrated an improved selection procedure that produced truly pluripotent cells able to be regrown into viable mice. Later that year, the same process was replicated successfully on human cells (minus the regrowing of a new human). The process had now proven the remarkable backward programmability of cells, but the resulting cells were not the sort of thing you could safely use in human therapies because two of the four magic genes are sometimes implicated in cancers. Cancers are cells that reproduce way too often (a bad thing), and stem cells are cells that reproduce more often than normal cells, but just often enough (a good thing). The four magic genes ramp up your reproductive rate to that of stem cells. But by retrofitting extra copies of these genes into the cell, you cause them to be overexpressed. And indeed, 20% of the artificially developed mice developed cancer. The next year, researchers discovered how to implant the four genes without retroviruses, leaving no genetic side effects from the virus itself. By 2010, researchers had figured out how to mimic the effects of the four genes artificially by synthesizing their corresponding messenger RNAs (mRNA) – the messages genes send to the cell's protein factories to express their proteins. These synthetic mRNAs could be introduced into the cell just during the reprogramming phase, without touching the original genome, and without the side effects of viral vectors, thus bypassing the cancer problems involved in overexpressing the four magic genes. In four short years, we went from wondering what would happen if we messed with biological time to having a full-fledged technology for safely reverse engineering almost any of the 37 trillion cells in your body back to an embryonic state and then forward engineering to any cell we please – including another whole version of you.

Regenerative Medicine

To get from your first cell to the 220 different types of cells that make up your final 37 trillion, some cells have to change their types along the way. These are the stem cells, named after the successively branching stems of a tree diagram that depicts the order of cell-type lineages. Some of these intermediate forms are dedicated to the one-time process of growing you up from an embryo. Their job is to get you there, and they aren't really needed after that. Another class of them, known as adult stem cells, hangs around in specific final tissues for occasional acts of redevelopment, either in tissues like skin or blood that need constant regeneration, or in tissues that don't normally turn over, but occasionally need some of their cells replaced due to wear and tear or injury. You have small reservoirs of these partially differentiated, adult stem cells in your brain, bone marrow, blood, blood vessels, muscles, skin, gut, teeth, liver, ovaries and testes (that we know of so far). These are the cells that account for multicelled life's ability to resist creeping entropy. Orderly things naturally wear out and break over time, so these perpetual keepers of the order keep making replacement parts.

Recall that the bane of complex, multicellular existence is cancer. In order to have 37 trillion cells functioning together in exquisite harmony, you cannot afford to have most of these guys reproducing on their own schedule. Cancer happens when some of them go off the reservation. So the complete animal needs to strictly control cell division. Most types of adult cells don't divide. They stay in their proper place in the tissue and maintain the collective order. The adult stem cells, which are relatively rare in number in the tissue, are in charge of the limited, controlled growing. They can both reproduce another undifferentiated version of themselves, and a more differentiated version of the kind of tissue that needs some replacing. So they have a limited license to reproduce, usually when the surrounding cells signal them that something needs replacing. The original embryonic stem cells, on the other hand, are designed for fast reproduction and differentiation, following an animal-building plan encoded in the genome. The equivalent of cancer here is birth defects. Someone doesn't quite follow the script properly.

The idea of regenerative medicine, of manipulating stem cells to repair or regrow damaged or diseased tissues, has been around since stem cells were first discovered, and has been used for some time in bone marrow transplants. What has limited its use until recently is this difference in the reproductive strategy between embryonic and adult stem cells. Some animals, like salamanders, have stem cells in an in-between state that can regrow an entire appendage on demand, firing up the original development program temporarily, then shutting it down again. Humans don't have this. We are programmed to build the whole thing once with the embryonic stem cells, then regenerate much more modest pieces of the body later with the adult stem cells. It is hard to extract adult stem cells from tissue (there aren't many of them), and because of their conservative reproduction rate, and sensitivity to the signals of surrounding cells, it is hard to culture them and get them to multiply on a lab plate. Embryonic stem cells are much easier to harvest, but since you have to get them from embryos, there aren't many donors. They multiply like crazy in culture, which is good, because one line of them can be used to regenerate a virtually unlimited supply. But it is much harder to get them to behave like the tissue you want to replace when you transplant them into a body, because they are naturally programmed for original development. Even on a lab plate, if they get together they start to form the three basic precursor tissue types of an embryo. When implanted, they often form a *teratoma*, a benign tumor composed of a jumble of tissue types.

There is an additional problem with using embryonic stem cells, because the "donating" embryo is a different individual than the mature patient getting a transplant, so the cells risk rejection by the patient's immune system. To get around this, the potential regenerative therapy has to pluck out the nucleus of the embryonic cell and replace it with a nucleus from the patient's own genome. This creates an initial hybrid: a cytoplasm from the embryonic cell containing the factors for stem-cell-hood, and a genome marked up by epigenetic factors from the patient for a mature cell type. iPS cells solve both of these problems. The patient has potentially 37 trillion donor sites, many of these in painless and expendable areas like the skin, and the eventually reverse-engineered iPS cell already has the patient's own genome in it. The pluripotent state can be used to make many clones on the lab plate, then forward engineering puts the stem cell closer to the target cell type to ease the risk of teratomas.

During the initial excitement about biological alchemy, when medical gold could only be manufactured using cancer causing genes and viruses, it immediately occurred to researchers that these iPS cells would still be useful in modeling the progression of disease in individual patients. Neurodegenerative diseases, like Parkinson's and Alzheimer's, often aren't apparent in a patient until the disease has progressed to a state where successful intervention is too late. It's not particularly feasible, or safe, to extract a section of a patient's brain to study the presence and progression of these diseases in neurons in the lab. But it is easy to take a skin cell or two, from individuals known to be at genetic risk for the disease, which will have the defective genes in their nuclei, reverse engineer them to iPS form, then forward engineer them to neurons. Now you have the disease – customized to a particular individual – modeled in cells in the lab (as many as you need) that you can study for disease progression and experiment on with targeted drug therapies to see what might work. You don't have to be as conservative with drug candidates because you only risk killing a few cells and not the whole patient.

This individual-model-of-disease application would have been revolutionary, even if iPS cells couldn't safely be put back into patients. But with the discovery of the mRNA induction process that bypasses the cancer and virus problems, suddenly we have an unlimited supply of whatever kinds of cells or stem cells we need to fix any condition requiring regeneration, derived from a patient's own genome. Researcher

s are trained to be cautious in their aspirations for what clinical applications a new discovery might lead to, but given that the previous laws of biological time have just been suspended, imagination is now justified. One application that already appears to be feasible is the regeneration of whole organs. This has already been done with a mouse liver. A small version was grown from stem cells in the lab then transplanted into a live mouse, where it matured into an adult liver with all of the appropriate connections to the rest of the body. Scientists started here because the liver is a maximally amorphous organ without a fixed structural shape. Portions of donor livers had already been used to grow new livers in recipients. The process is harder for organs like the heart, which have a very complex architecture and several cell types. But a process that seems to work is to take an existing, fully formed heart from a cadaver, immerse it in a solution that dissolves all of the original cells leaving just the collagen and extracellular matrix, then seed the form with the recipient's reengineered stem cells. With the prototype structure to guide them, the developing cells figure out where they need to go and what they need to become. Voilà! The recipient now has a new heart, custom built from his own genome, for transplant. An even more imaginative technology developed recently uses a 3-D printer to build the collagen scaffolding from a scan of the geometry of the recipient's original version of the organ. Now you have your own genome, and your own shape as well.

A Cell is a Cell is a Cell

The repeal of the law of biological time tells us something that we didn't expect about cells. Although single-celled eukaryotes sometimes form colonies of like individuals, the most common community form is the multicelled organism in which cells differentiate into the specialized cogs of a larger clockwork. Most prokaryotes, like bacteria, associate with each other in larger colonies in which the individuals retain their autonomous identities. These colonial associations exhibit many of the same communal mechanisms for inter-cell signaling, and adhesion, and self-organization in the building of structures. But we thought for sure that these were two fundamentally different kinds of phenomena. Communal prokaryotes can always go their own way, if they choose. Eukaryotes in multicelled bodies, on the other hand, sure looked like one-way bio-bricks – cells that had permanently sacrificed their individual identities and given up all of their potential futures for the common cause. There are no separate existences. You can't leave the colony.

We've known of exceptions to this rule, but they are mostly marginal or primitive. In humans, for instance, two of the final cell types retain their independent existence with the intent of leaving the colony – the *gametes*: sperm and ovum. Sperm cells leave the body all the time in order to start an independent life, and the ova eventually leave too, incorporated into the next generation. At the lower end of the animal kingdom we have sponges, which differentiate from a single germline cell into a multicellular body composed of four distinct cell types. An inner layer of cells, called *choanocytes*, are strikingly similar, visually and genetically, to single-celled eukaryotes, called *choanoflagellates*, which sometimes form small colonies. Sponges are currently regarded as the most ancient of the surviving multicelled animal forms, and their primeval status is evidenced by a peculiar flexibility that they retain from their unicellular ancestors. Even though the body is built by the division of labor into four cell types, they are all germline cells. Any one of them can be detached from the body and go on to reproduce a whole new sponge. These cells also have not completely forgotten their primordial community-by-convergence programs. If you artificially homogenize a sponge by breaking it into a pile of single cells, the cells will seek each other out and re-associate into the original multicelled form. So sponge cells have learned how to reconstruct their multicelled parents through a process of controlled elaboration, but they still retain vestiges of an evolutionary past when they used to have to find each other from scratch.

But we thought we lost this flexibility as animals got more complex. In many plants, you can take a cutting off the parent, plant it somewhere else, and it grows into a whole new parent again. It has long been an accepted dogma of biology that you can't regrow humans from cuttings. We were wrong. Apparently you can. The designated reproducing cells, sperm and ovum, are just one of the ways to do it. Any cell has the potential, it seems, even though the backward pathway appears to be entirely artificial. So now we have to rethink this. It seems as though a cell is a cell is a cell, regardless of whether it makes its living as a jack-of-all-trades prokaryote or as a very specialized cog-in-the-machine eukaryote. Each cell is a potential individual that never really loses its identity or its memory. All of the observed cell roles (so far) are not driven by some fixed biological fate, but by the accidents of their current organization. This does seem like a very versatile way to design life. And it seems to reinforce the notion that single cells are the reference standard for life in any form. A remaining mystery is this: if multicelled development is so modularly robust, in any direction, why don't some of these other trajectories happen naturally in multicelled animals? Or perhaps have we just been misinterpreting what happens during embryogenesis when development first reaches a point, like the pharyngula, and then backtracks to something else?

How does normal development keep all of these backward pathways in check if they are actually possible?

When Personhood Begins

iPS cells were first discovered during George W. Bush's 7-year ban on federal funding for embryonic stem cell research (except for the 12 ES lines developed prior to 2001). The ban was put in place to preserve the sanctity of 5-day-old blastocysts left over from in vitro fertilization clinics that would be destroyed anyway – so its net effect was simply the suppression of research. Those leftover embryonic stem cells were deemed to be "human life," presumably because they already had souls at this point, even though they had not gone through the traditional ensoulment route of natural conception. It is ironic that when they first emerged, iPS cells were hailed as the right and proper way to obtain stem cells for research, because they did not derive from embryos. 'Pluripotent' somehow seemed less sanctified than 'embryonic,' even though they were exactly the same kinds of cells, and could be used to create the same kind of blastocysts. This being a fairly novel route to new humans, it apparently didn't occur to the policy makers that ensoulment might be involved here – so these weren't "human life," they were just cells. After all, they had been just ordinary body cells moments before their pluripotency was induced. Sometimes it's a good thing when politicians don't understand science.

We have been dragging the "life begins at conception" idea along with us through the ages, trying to retrofit it onto each new discovery about how reproduction actually occurs. It is a concept based on the original misconception that complete humans spring forth in a single event of spontaneous creation. It's much easier to imagine ensoulment in that context. But we have long since known that mammalian life doesn't happen that way. The medieval homunculus was the first creative attempt to have it both ways. A zygote, or an embryo, or whatever it was that was happening early on in the uterus, was thought to be just a miniature version of a whole human. So it gets a complete soul right from the start. It's just the body that gets bigger.

Eventually, however, we got a good look at these early-stage embryos, and they sure didn't look like little humans. They didn't really look like anything at all, just undifferentiated cell masses. Life became even more of a mystery, and we lost confidence in our ability to locate a convincing phase in this development for when ensoulment occurs. We still had a canonical event – conception – so whatever biological thing gets created at conception became the canonical form of a new human life. There was no other way to create human life, so at least we could zero in on a particular place and time. When in vitro fertilization (IVF) came along, there was no longer a canonical place and time for the conception/creation event. And there was quite a bit of angst about whether humans were usurping the power of Gods with this technology. But it did seem to work. We certainly couldn't say these "artificial" children did not have souls, so we had to get used to artificial routes to ensoulment.

But at least there was still a canonical genetic event – when two haploid cells (sperm and ovum) fuse to form a diploid one (zygote). We call this *meiosis* – sexual recombination. It might now also happen in a test tube, but this could still be pointed to as the point of human ensoulment. If you had been paying close attention to the biology, which most people weren't, you would already have seen some problems with even the notion of there being a unique *time* for ensoulment. The original zygote develops over 5 days into a blastocyst, a small mass of embryonic stem cells surrounded by a sphere of cells that are the precursors to the eventual placenta. By day 8-9, the blastocyst implants itself into the uterine wall and differentiated development begins. Like almost all processes in nature, this one is probabilistic. Sometimes a zygote completes the process, sometimes not. It wasn't until we intervened in the process with IVF that we had a good way to calculate this probability. It turns out that from 30% to 70% of zygotes fail to implant. Even after implantation, another 25% are lost by the 6th week. That's why IVF clinics make so many copies of the embryo, and then have a bunch left over. There is trial and error involved. You certainly can't blame the flushing of these zygotes and ES cells on the mother, in the case of natural conception. They are "acts of God" in insurance lingo. But this means that if Gods are ensouling the zygote during meiosis, they are also wantonly destroying human life, just like IVF clinics and stem cell researchers.

Next came whole animal cloning. There are two parts to a zygote that make it capable of generating a human life: the DNA describing the final individual, and the surrounding cytoplasmic state of the larger cell that causes the development genes to be turned on. Researchers discovered that you can take the nucleus (DNA) out of a normal body cell, and use it to replace the original nucleus of an ovum (unfertilized egg). The ovum contains the right protein state to start the development genes of the foreign DNA. This causes a second copy of the animal, whose DNA it is, to develop. So which part is the potential human: the DNA or the surrounding cell state that says, "start developing"? This ambiguity was exploited by some early researchers to deftly get around prohibitions on research on human embryos. By replacing the DNA in a cow ovum with that of a human, you obtain a hybrid zygote that is not strictly human – at least the embryonic part isn't. This is the kind of rabbit hole you go down when you try to force fit characteristics of whole animals onto molecular biology.

The discovery of iPS cells now pretty much finishes off this whole conceptual quagmire. It turns out that there is nothing unique, or isolated, or canonical, or rare about cells that have the potential to develop into whole human beings. Almost any human cell can. There are many ways in which this can be done. Secular society has already gotten used to the idea of separating the concept of a person from the concept of the first cell that leads to the development of that person. By legalizing abortion, civil societies have had to come up with criteria for when enough of the characteristics of a person are possessed by the fetus to call it a person, to assign it rights and legal protections, and to prohibit its willful termination. This is not an easy task, given the smooth continuity of embryonic development. But the criteria we often choose – a beating heart, a developed nervous system, the ability to feel pain – are the kind of things we typically assign to whole persons, not single cells. We have empathy for living things that exhibit these properties. We don't have empathy for single cells, because they have none of these features.

It's hard to imagine that the "life begins at conception" folks have any real empathy for single cells either. They have learned to transfer their reverence for whole humans to what they thought were a rare and distinguished class of single cells – the cells with the *potential* for developing into whole humans. This is what distinguishes zygotes from bacterial cells, or skin cells, or amoeba cells – a very uncharismatic, microscopic bag of chemicals, the whole lot of them. But they are still just single cells. It's hard to have any natural feelings for features like *potential* in a cell. It's no accident that the pictures protesters display outside of abortion clinics are of relatively late term fetuses. It wouldn't be very compelling to show pictures of zygotes. No one thinks "person" when they see these. But this belief in the sanctity of potential life, or life with potential, can lead true believers to advocate draconian laws that apply norms of person-to-person conduct to person-to-cell conduct – banning life-saving research, and holding doctors and mothers liable for murder. They are willing to sacrifice the lives of actual persons we know and love to preserve the lives of cells that we can't possibly know or love. Well now look what's happened. Suddenly you discover that you are composed of some 37 trillion of those potential humans. They are *all* unborn children. When you shower, or wash your hair, you are killing thousands of them. Here's an even creepier thought. One recent, very efficient protocol for harvesting adult cells for iPS transformation is to collect the naturally exfoliated kidney cells in a patient's urine. Oh no! Every time you urinate, you are peeing away human life!

So what's a responsible and appropriately reverent human to do now? Stop bathing? Stop peeing? It's really not that complicated. You just have to let go of some rather abstract notions about life that no longer make sense. Life doesn't begin at conception. That's meiosis. Meiosis results in a net reduction of life. Two lives that *already* exist cease to exist and become a single, fused life. After *that,* new lives start to begin – by *mitosis,* nature's more ubiquitous reproduction method, the division of one life into two lives. Then four, then eight, then eventually 37 trillion. Personhood doesn't begin at the first cell that has the potential to become a person. The *development* of a person begins at the first cell. Every one of these new lives is potentially a person. They die all the time through normal acts of God and normal acts of persons. This death is an integral part of life. It is ethically, and emotionally, neutral.

There are many paths to realizing any cell's potential personhood. Only the ones that manage to get a certain distance into development, along any path, actually become persons. All of these paths involve the gradual emergence of a person to which ethics and laws and emotions eventually attach. Just as with space aliens, the holy books don't have anything to say about in vitro fertilization and induced pluripotency (or meiosis and uterine implantation). These would have been hard to anticipate, so they are exercises left to the reader. If you believe that Gods create and ensoul humans behind the scenes, nothing really changes. We are simply discovering that almost every biological process is based on continuous change rather than sudden phase shifts. So we have to deal with the gradual emergence of things, like personhood. Certainly Gods can handle this, and put souls in when the time is right.

As with life itself, persons don't happen suddenly; they fade in. We struggle with this emergence in the context of human embryology, but there are similar continuums to deal with in the evolution of humans from non-humans. Modern humanhood emerged one mutation at a time, so things like ethics and rights and responsibilities and dignity emerged gradually as well. We don't face any present close calls with our hominin ancestors because they are all extinct, but we are only about 6 million years removed from chimps and bonobos. They are functionally (and morphologically) pretty close to us, so we have to consider the very real possibility of their rights and responsibilities and dignity. This continuum occurs all the way down the animal kingdom of present day species until you reach sponges. It's not easy being human.

10 | Another Diversion

Which came first, the chicken or the egg?

The egg.

We are now in a position to answer this ancient question. Like many similar paradoxes, it is based on the notion of an infinite regress. We observe some alternating sequence of A causes B and B causes A, and then try to back it up to a first A or B. How can there be such? You could say that the Biblical answer is the chicken, since God reportedly created the first animals de novo and told them to be fruitful and multiply. But the story is lacking in details. It doesn't specifically address chickens or the arcane minutiae of the creation process. It's compatible with God having chosen the egg first. All it implies is that there was a first chicken and a first egg.

From the biological perspective, we are able to get an objective fix on the first chicken and the first egg because we can view a smooth continuum of evolution from some point before there were either chickens or chicken eggs. Evolutionary changes occur one mutation at a time, so as with so many biological categories it is more appropriate to say that chickens and their eggs gradually *emerged* from pre-chickens over some extended period of time. But we can approximate an answer by picking some one mutation in the series that made the net difference – the pre-chicken gene that first mutated into a full chicken gene. This would have happened in either or both of the pre-rooster or pre-hen parents of the first chicken. When the pre-hen's embryo was developing, one of her *oocytes* (egg cells) would have sustained the appropriate copying error in its genome. This only affected the oocyte, not the rest of her whole pre-hen body. Mutations in the non-germline cells are not passed on. Similarly, when the pre-rooster's germline stem cells divided to produce sperm cells, one of them sustained the appropriate copy error. Again, this didn't affect the pre-rooster, just one of his sperm. When the pair mated, if either the sperm or the ovum had the new gene, the resulting zygote would develop into the first chicken egg. We wouldn't have the first chicken until the first chicken egg finished embryogenesis. So the egg preceded the chicken.

If you want to be pedantic and say that chickenhood couldn't possibly have been defined by one mutation, we can always generalize and say that the *emergence* of the first chicken, over time, started with the first mutation in an egg and ended when the last mutation reached a chicken. Another way to put it: non-chickens can produce chicken eggs (because this is where heritable mutation happens), but non-chicken eggs can only produce non-chickens (because they just elaborate the genome they are given).

11 | Death

To every thing there is a season, and a time to every purpose under the heaven: A time to be born, and a time to die.

> – Ecclesiastes

A. DEATH IS CERTAIN
1. There is no possible way to escape death. No-one ever has, not even Jesus, Buddha, etc. Of the current world population of over 5 billion people, almost none will be alive in 100 years time.
2. Life has a definite, inflexible limit and each moment brings us closer to the finality of this life. We are dying from the moment we are born.

> – from an Internet version of the Tibetan Buddhist nine-round death meditation

... in this world nothing can be said to be certain, except death and taxes.

> – Benjamin Franklin

One foundational aspect of the meaning of life that can be found in every human tradition, in every age, is that it has an inherent end stage – death. There is significant disagreement about what happens after that, but everyone seems to agree that the body will die. Ashes to ashes, dust to dust. It's inevitable. We rage at the unfairness of death when it takes someone unexpectedly in their youth, but we wax poetic about the eventual winding down of old age. We can see this kind of death coming from a long way off. We reflect on how it is an integral part of life. One of the great rites of passage for a human is to come to grips with this realization. The open-ended plans and limitless possibilities of youth eventually have to be reconciled with our finite existence.

Religious views of life are essentially built around the concept of death. What you should or shouldn't do in this life is always defined relative to what consequences it will have for you at death. The particular codes of conduct, the nature of the sacraments, the scripts for the rituals, who's a saint, who's a prophet, what's a miracle, all come and go, but you can't alter the basic framework of birth, death, then whatever happens after death. Each religion defines itself by its integrated story of these three phases. Death has to be in the middle – that's what religion is for.

You don't need religion, of course, to persuade you of the inevitability of death. This comes to us naturally. Even if you don't see a lot of dying, you see a lot of aging. Everything grows old, weathers, declines, atrophies, wears out, dries up, wastes away, decomposes over time. This is our folk theory of entropy. Bodies are no exception. In biology, this process is called *senescence* – a gradual winding down of biological structure ending in disintegration. If bodies didn't senesce, they still wouldn't necessarily be immortal, because you can always be killed by accidents, or violence, or infectious diseases. But if you manage to escape all of these precipitous endings, senescence will get you eventually. As biology has progressed it has improved our understanding of senescence. We have learned a few things about the inevitability of this kind of death that may surprise you – among them, that it's not inevitable. Some organisms don't senesce. This is not new knowledge. We've known for some time that death is a relatively recent invention in evolutionary time. Yes, that's right. Death evolved. It's not an essential feature of life; it's just a common feature of complex creatures. As we understand more about why complex creatures senesce, and why some of them apparently don't, it's looking increasingly likely that we will be able to reengineer this process in humans. Biology can't really help us with the inevitability of taxes, or what comes after death, but it may allow us to achieve a certain measure of immortality.

Biological Entropy

For a long time, we thought that single-celled organisms that subdivide by mitosis, like bacteria, were effectively immortal. Bacteria die all the time, of course, in huge numbers, but not generally because they wear out. They get killed by outside forces. No one had ever observed senescence in bacteria, and it made sense that this might be an unexpected phenomenon because they are pretty simple, their metabolisms repair damage, and they are constantly creating new copies of themselves which gives them a chance to dilute any accumulated damage by one half. The problem with observing any property of an individual bacterium over time, though, is that it is hard to keep track of them. It wasn't until recent technology enabled this individual tracking that we learned that individual bacteria can senesce. Subdivision by mitosis doesn't always distribute the accumulated entropy evenly. One of the daughter cells gets less, giving it a better restarting point, but leaving the other a little more worn out. If the same individual is on the losing side of this transaction often enough, it will senesce. In another sense, though, this does allow an individual genome to be immortal because each daughter cell is a clone of the original, so the descendant line that keeps getting the better half keeps that genome going indefinitely. And the prolific reproduction rate means you can't kill them all by outside forces, so they are truly immortal. This is impressive until you realize that bacteria mutate and swap their genes so often that even clones aren't true clones for very long. *Something* survives forever, but it's hard to relate that something to our more stable notions of individual identity.

This effective lack of senescence stands in contrast to single-celled eukaryotes, which do exhibit some of it, to plants, which exhibit more of it, and to animals, which exhibit a lot of it. Senescence seems to vary with complexity, and this makes intuitive sense. The more interdependent parts you have, the more points of systematic failure you have. You are constrained in your ability to replace or renew all of the parts in a coordinated manner, so some parts will wear out. The odds are against you in the long run. This observation raises the possibility that senescence evolved not by selecting for timeout-death as a feature, but simply because the evolution of complexity itself carried with it the burden of greater accumulated entropy, making senescence a side effect. Still, there is a lot of variation, from species to species, in how soon senescence begins, if at all, and this variation doesn't seem to correlate well with complexity.

So early theorists looked for some sort of biophysical law that would explain the huge variation in lifespan among different species. One that held up for a while over many of the observed lifespans at the time was the "rate of living" hypothesis. Small, fast creatures, like mice, live no more than 3 years, where large, slow creatures, like elephants, can last for 70 years. It is known that the normal metabolic activity of cells creates some oxidative damage as a side effect, so the idea was that animals, like cars, have a certain amount of total mileage in them. If you race your car every day, it will not last as long as it would if you only drove it slowly on Sundays. This handled the elephant-mouse dichotomy, but it broke down as more varieties of lifespans got added to the mix.

We still don't have any overall biophysical explanation for the huge disparity in animal lifetimes, so it seems that evolution is responsible for this on a species-by-species basis. But because of all the recent research on human senescence, we have a pretty good understanding of the kinds of things that cause our bodies to wear out over time. The base cause is oxidative damage. For eukaryotes, oxygen is both our friend and our enemy. We need it, coming through the blood from the lungs in the stable form of O_2. This is what the cell's mitochondria use to produce the cell's energy. But one of the side effects of this energy production is various reactive oxygen species, molecules in which the embedded oxygen can easily react with surrounding molecules, such as DNA, proteins and lipids. These reactions damage the structure of DNA, causing coding mistakes, damage the structure of proteins, altering or destroying their function, and damage the integrity of lipid structures. All of these are not good for the cell, as you can imagine. One of the benefits of the eukaryotic cell design is that it isolates the oxygen metabolism inside the mitochondria, and further insulates the DNA inside the nucleus. But as with steam engines, there are always leaks. So carrying on oxygen-based life at all carries with it an exposure to a small amount of structural damage. The longer you live, the more this will accumulate.

As impaired components gradually degrade the cell's normal operation, the cell is less efficient at cleaning up accumulated molecular junk both inside the cell itself, and in the spaces between neighboring cells. Structures such as blood vessels and the collagen proteins in skin become less flexible. Cells eventually lose their ability to divide, and thus their embedded mitochondria, which manage to dilute their own oxidative damage by continued subdivision, must cease to divide as well leading to their degeneration. It just keeps going downhill until eventually you reach whole organ failure.

The Evolution of Death

If senescence were as simple as this creeping, oxygen-based entropy, there would be no immortal eukaryotic species. But there are. Plants in particular, which are eukaryotes and face oxidative damage both from photosynthesis in their plasmids and energy production in their mitochondria, often don't senesce, or only senesce in certain expendable parts of their structure. They die from parasites, or environmental damage, or habitat change, or competition for resources from other nearby plants, but in general they just keep growing larger until some outside force stops them. The bristlecone pine in the American West has been known to survive for about 5000 years (so far). Animals at the lower end of the kingdom, such as hydras and jellyfish apparently don't senesce. Lobsters are also believed not to senesce. Their reproductive prowess actually increases as they grow older.

Also, if senescence were just a critical accumulation of oxidative damage, there would be a direct correlation between how much oxygen you burn and how long you live. But there isn't. Species large and small, simple and complex live for all manner of short and long time periods, burning oxygen at very different rates. This independence of lifespan from the rate of metabolic damage is due to the fact that cells and organisms have evolved all kinds of elaborate mechanisms for the continuous repair of all of this damage. We don't suffer entropy lightly. We do something about it.

There are mechanisms for repairing DNA damage as it happens. In humans, for example, there are more than 10,000 oxidative hits on DNA per cell per day. The normal repair mechanisms in our cells, though, are more than capable of keeping up. There are mechanisms for breaking up the leftover molecular junk and clearing it out of the cell. There are mechanisms for continuously repairing the lipid structure. There are special chaperone proteins that help get other proteins to their rightful places, that help them fold into their proper shape, that help them undo dysfunctional folds, or attachments. There are stem cells that replace cells that have taken too many hits to continue. There are phages that destroy cells that have gone off the reservation. There is a process called *apoptosis*, by which a damaged cell will commit suicide, either because it senses its own damage, or because of instructions from neighboring cells. There are "heat shock proteins" that respond to environmental stress (of more than just heat) to put whole subsets of these repair mechanisms into higher gear for a period of time.

With all of these Mr. Fix-it routines, you may be wondering why we aren't *all* immortal. Why does any species senesce? Well it's not because different species have only a few of these mechanisms. They are all highly conserved across the animal kingdom. We all have most of them. So this doesn't account for the variable lifetimes of different species either. The naked mole rat, for instance, a mouse-sized rodent that lives underground, lasts for at least 17 years in the wild and twice as long in captivity, yet the common house mouse lives less than a year in the wild and only 2-3 in captivity. The little brown bat lives at least 34 years in the wild, yet its close cousin, the evening bat, lives at most 6 years in the wild. A distant cousin of ours, but still a primate, the common marmoset, lives 7 years on average, but no more than 16. Similar organisms, similar levels of complexity, similar complement of repair mechanisms. Why the difference?

The only apparent explanation is that evolution *turns off* the repair functions at different rates in different species. Humans have a period, for instance, between approximately 20 and 40 when we have finished our development program, but remain homeostatic, fixing everything as it breaks. Then things start to slowly fall apart. Why couldn't we just keep that middle program going forever? If we *evolved* this particular timeout switch, you might wonder, what possible selective value could that have had? The consensus of biologists on this matter, over many years, is that each species' present rate of senescence onset is a function of its rate of natural survival in the environmental conditions under which it evolved. The only reason for evolution to select *for* mechanisms that get you to a certain age is if you are still reproducing at that age. If you are not, and don't have some other compensating characteristic that makes the survival of your existing children more probable, there is no selective pressure to keep you alive. This requires evolution to get you through at least your reproductive years if you are going to be a surviving species at all. But if the natural rate of attrition for your members, through starvation, predation, or any of the other environmentally driven death scenarios, pretty much kills you all off by age N, then none of your "past N traits" matter. They never get a chance to prove their worth.

Evolution is not so much selecting *for* senescence after a certain age, as *failing* to select for repair after a certain age. Whatever random mutations break the repair mechanisms after a certain age face no negative selective pressure. They don't matter then. So these mutations will accumulate. Mutations that break repair before the magic age will cause you to leave fewer progeny than your fellows, so your unfortunate mutations will be weeded out of the gene pool. This helps explain some otherwise anomalous differences between species. Birds for instance, have significantly longer lifespans than many of their similarly sized mammalian counterparts because their airborne lifestyle makes them less prone to predation. So a bird with a mutation that pushes the senescence onset out a little further has a chance to survive long enough to reproduce more offspring with that trait. A mouse in the wild, on the other hand, that gets a lucky mutation extending its senescence age out to 5 years will die before it reproduces any more progeny than its fellows who have the normal rate. This keeps the trait alive temporarily in its descendants, but there is nothing to protect that trait from getting broken by re-mutation, because none of the progeny will get that far.

It's also possible for evolution to actually select *for* life-ending characteristics later in life. This is called *antagonistic pleiotropy*. Pleiotropic genes are ones whose expression causes more than one final trait in the organism. They just happen to be tied together by the downstream molecular machinery. If such a gene were to have a beneficial effect early in life and a harmful one later in life (that's the antagonistic part), it would be selected for during the organism's reproductive years. The harmful later effect would be outside the scope of heritability. It is protected from re-mutation by the constant selective pressure on the early beneficial trait.

Clocks and Switches

In order for evolution to be sensitive to lifespan-dependent traits, it seems there must be some kind of biological clock in organisms that can serve as a target for selection. This could be a single structure that counts off units of biological aging for the whole animal, or perhaps multiple such structures whose rates of change are tied to the various rates of biological aging in different organs. It could even be an entire system of changing gene expression and differentiation that reaches some natural end point. The development program for most animals is a structure like this. There isn't an abstract clock that counts down the time allotted for development, rather development proceeds until something is finished. The process may take varying amounts of time to complete for different individuals, but the completion state marks the end of a lifespan stage.

A complex developmental clock of this last kind seems unlikely in the case of closely related species with vastly different lifespans – like the two bats, or the two rodents. They have very similar development programs. Evolution must have picked on something much simpler and more sensitive to small changes. For a while, the biological community thought it had found such a universal clock. Leonard Hayflick conducted experiments in the early '60s to determine how many times normal human body cells would continue to divide in a lab culture. It turned out there is a fixed limit: 40-60 times. After that, the cell will enter a senescent phase and cease dividing. This became known as the *Hayflick limit*. Later, the molecular mechanism that enforces this limit was discovered in the form of *telomeres*. These are small, successive, repeating sequences of DNA at the ends of chromosomes. When a chromosome is replicated during cell division, the replication machinery cannot get all the way to the end of the chromosome without doing some damage to whatever is at the end. So these successive repeats are expendable pieces of DNA that can absorb the end damage, rather than having it occur to an ending gene sequence. So each time the cell replicates, the telomeres at the end get a little shorter. When a cell runs out of telomeres, it shuts down replication in order not to damage real genes.

The Hayflick limit for other species appears to correlate well with their expected lifespans, so we appear to have a stunningly simple biological clock, and a DNA implementation of it, via telomere length, that is eminently evolvable. It is easy to imagine how small copy errors in replication, which duplicate or delete telomere repeats, would lengthen or shorten the timeout mechanism. An evolvable clock! Would that it were that simple. But very little in biology turns out to be that simple. Although there is a correlation between original telomere length and lifespan, and between aging and shortening telomeres, the starting telomere length is only a suggestion for time-to-senescence, not a hard limit. Some fully differentiated cells like neurons and heart muscle cells don't divide at all. Others, like stem cells, divide indefinitely. The organism achieves this variable rate of replication with an enzyme called *telomerase*. Telomerase can add DNA repeats back on to the telomeres, extending the timeout clock. Stem cells express telomerase to keep themselves effectively immortal. Differentiated cells don't, which makes them finite.

Cancer cells divide indefinitely as well. It has long been thought that the telomere length timeout mechanism first evolved as a hedge against cancer, kind of like term limits for politicians, or expiration dates on credit cards. If a normal human cell can't divide more than about 50 times, then if one of them goes haywire and starts replicating out of turn, it will stop out at 50 before a tumor gets too big. This may actually function as a brake for some cancers, but the ones that go on to kill us have learned to keep themselves going by expressing – you guessed it – telomerase.

So what we really have is a default clock with a very flexible override switch. Evolution can fiddle with either the clock or the switch, or both. It appears now that telomere length is a reliable indicator of age, in that it decreases as we age, as well as an indicator of non-standard aging. Prematurely aged individuals have shorter telomeres and very long-lived humans have longer ones. Environmental factors can also manipulate the switch. Oxidative stress drastically shortens telomeres, and healthier lifestyles can lengthen them. So via the telomere clock, we can determine the current state of senescence in an individual, but with the ubiquitous availability of telomerase, and no doubt other tunable switches, we don't have a handle yet on the possibly many molecular causes.

The upshot of all of this is that death by senescence appears to be an evolved trait, not a necessary fact of life. If evolution can toggle the clocks and the switches to pick when, if at all, the repair mechanisms will shut down, perhaps we can too – extending their duration, or perhaps keeping them on indefinitely. Biology often frustrates us with massive side effects when we target one aspect of our physiology for intervention. Pharmaceutical companies know this. But every once in a while we discover some fundamental modularity in life that is very flexible, tunable, and malleable – like iPS cells. There is an opportunity to run the mechanism in a different order and it still works because we are reusing the body's subsystems in a way that mimics evolution's own repurposing of parts. When telomeres and telomerase were first discovered, it was speculated that telomerase itself might be the elegantly reusable switch to use. Just turn it on in normal cells and you'd get instant immortality! Well, this works, but it also removes the break on cancer. Cells replicate in their current, controlled order for a reason. Telomerase expression isn't a global switch for "continue to repair everything." It's the local mechanism, used in many different ways in different cells, to implement such a directive.

We haven't found such a master switch yet, but we've found some things that suggest there may be some. One recent, intriguing set of experiments reasoned that since all organs and tissues seem to begin senescence at about the same time there must be some global, coordinating signals. The body's main system for propagating global signals in real time is through the blood, since it circulates through all of the tissues. This is why the immune system operates in the blood. To test whether a "senesce/don't senesce" signal might be in the blood, the circulatory systems of an old and a young mouse were surgically grafted together. Sure enough, after 4 weeks, the heart of the old mouse regressed to the youthful state of the younger mouse. Later investigations isolated a particular protein in the blood that produces this effect. Expressing that protein independently in older mice had the same rejuvenating effects as the shared blood. This is the kind of switch that you want to find. It appears to be at least one of the body's own old/young master switches. It controls all of the intricate and coordinated downstream processes that implement continued repair. You can reuse the body's own mechanisms – whatever they are – to get this done in the proper way.

Even without a working knowledge of the body's master switches, the recent explosion in stem cell research, particularly that stimulated by iPS cells, is giving us a working knowledge of how adult stem cell types repair organs and tissues, either because of injury and disease, or because of senescence-related failure. Since we will be able to make unlimited supplies of these tissue-specific stem cells from a person's own differentiated cells, we can intervene in the senescence process by regenerating lost or senescing stem cells. This is more of a downstream intervention than global switches, and it is tissue specific, but it has the desired leverage of using the body's own rejuvenation strategy.

Dealing With Immortality

Given these recent discoveries, it now appears likely that in the near future we will have the technology to indefinitely extend human life – a kind of assisted immortality. A remaining question is: would this be a good idea? We already have the technology for cloning humans, but we don't use it, because it doesn't seem to solve any pressing problems. And it would certainly create some new problems. But immortality, or at least a little bit of life extension, is likely to appeal to a lot of people. Just look at what we do now through cosmetic surgery to preserve youthful appearance, and through extraordinary medical interventions in hospitals to keep people from finally dying. The temptation will be great.

It's really hard to imagine what all of the ramifications of this might be for our future, but there is a troubling lose-lose scenario. If we end death by natural causes, but don't stop reproducing new humans, our numbers will keep growing until we exhaust the available resources on the planet. This is not sustainable. If, on the other hand, we set up some kind of law (assuming we could get everyone to go along) where you can't reproduce a new human until one dies of unnatural causes, keeping the current population constant, we will no longer be able to evolve as a species. The gene pool will remain static. This is OK as long as the planet doesn't change, or some series of new pathogens doesn't evolve and sweep through our population because none of us has evolved a lucky mutation to survive them. We only have about 3.5 billion years left before our sun expands to extinguish all life on Earth, so we are toast anyway (literally) in the long run. But it could happen a lot sooner for humans if we can't evolve.

These are all practical considerations for future humans. This particular change in the meaning of life – the optional nature of death – will likely affect people living today, presenting us with new options we never anticipated, and perhaps didn't really want. But it's hard to underestimate how philosophically disruptive this particular change in the meaning of life will be for most religions. Until now, religions have skated through mostly untouched by the biological discoveries they didn't anticipate. Life on other planets, evolution from single cells, alternate ways to make humans – all of these are details that weren't explicitly addressed in the simpler religious world orders, so Gods can always be penciled in as behind-the-scenes orchestrators. Souls can pop in and pop out at whatever turn out to be the latest definitions of the major transitions of life. But what happens when you eliminate one of those transitions altogether? Death! This is pretty fundamental stuff. The inevitability of death is woven into the very fabric of religions. Souls will be stuck on Earth, unable to complete the rounds. The great wheel of life will grind to a halt. The orderly procession of souls to judgment day will slow to a trickle, limited to those that die unnaturally.

Perhaps the way to finesse this is to note that humans still can't realize true immortality. If we're not careful, some of us will get picked off by unnatural causes, and eventually the sun will get us all anyway. So we would just be delaying the inevitable – albeit for a long time. Maybe we will just be a temporary demographic bulge in the previously uniform procession of souls, like the baby boom generations.

12 | GMOs

Genetic engineering doesn't happen in nature. Scientists force genes from bacteria and viruses into plant DNA, which result in dangerous side effects.

– Institute for Responsible Technology website

What would you think if I said that your dinner resembles Frankenstein - an unnatural hodgepodge of alien ingredients? Fish genes are swimming in your tomato sauce, microscopic bacterial genes in your tortillas, and your veg curry has been spiked with viruses.

– Sara Chamberlain, *New Internationalist Magazine*

Do you know about *dihydrogen monoxide* (DHMO)? Probably not. Well, you should. It is a colorless, odorless chemical compound, sometimes simply referred to as *hydric acid*. Its basis is the highly reactive hydroxyl radical, a species shown to mutate DNA, denature proteins, disrupt cell membranes, and chemically alter critical neurotransmitters. The atomic components of DHMO are found in a number of caustic, explosive and poisonous compounds such as sulfuric acid, nitroglycerine and ethyl alcohol. Accidental inhalation, even in small quantities, can cause death. Prolonged exposure to solid DHMO causes severe tissue damage. In gaseous form, DHMO can cause severe burns. Although it is sometimes used as an industrial solvent and coolant, it might surprise you that it is also used as an additive in food products, including jarred baby food and baby formula, and even in many soups, carbonated beverages and supposedly "all-natural" fruit juices. One of its residues has already been found in human blood. Because of its reactive properties, it is hard to predict the downstream molecular changes when it is added to solutions. And once it gets into the environment, it stays there.

There is no feasible way to get it all back out. The industrial chemical giant Monsanto uses it in many of its products. It is now present in a large percentage of the US food supply, but US regulators do not require it to be disclosed in food labeling. Most companies that include it as an ingredient don't disclose this. Although scientists and government regulatory bodies insist that it is safe, ordinary consumers, when they are confronted with the properties of DHMO enumerated above, are concerned. Various groups have signed petitions to have it banned, but politicians seem to show no interest. There are many documented studies showing the adverse health effects of compounds of which it is a part, though none (yet) that implicate DHMO directly. But why take chances? To be safe, savvy consumers should just avoid it altogether.

You might be familiar with DHMO under its more common chemical symbol: H_2O.

Do you know about *genetically modified organisms* (GMOs)? Well, you should. If you pay attention to food safety or food labeling, it is hard to miss them. The Internet is full of sites that will tell you of their dangers. These are some of the things you can find there. They are virtually indistinguishable from non-GMOs, both in whole form and as food additives. Genes from GMO plants can actually cross-pollinate with other plants. Popular GMOs promoted by large seed companies can lead to monocultures that reduce natural genetic variety. GMOs with survival or reproductive advantages over other organisms can become invasive species, crowding out weaker competitors. One of the residues found in some GMO foods was once found in human blood. Because new strains of GMOs involve genetic changes, it's hard to predict the downstream molecular changes that will result. And once these new genes enter the environment, they become part of the surrounding gene pool. There's no feasible way to get them all back. Monsanto also uses GMOs in several of its products. GMOs are now present in a large percentage of the US food supply, but US regulators do not require them to be disclosed in food labeling. Most companies that include them as an ingredient don't disclose this. Although scientists and government regulatory bodies insist that they are safe, ordinary consumers, when they are confronted with the properties of GMOs enumerated above, are concerned. Various groups have signed petitions to have them banned, but politicians seem to show no interest. There are many documented studies showing the adverse health effects of compounds of which GMO ingredients are a part, though none (yet) that implicate GMOs directly. But why take chances? To be safe, savvy consumers should just avoid them altogether.

You might be familiar with GMOs under their more common name: *life*.

This chapter takes a somewhat different tack than previous ones. We have been looking at some of the new meanings of life, new concepts that alter our traditional understanding of life for which we often don't yet have new names. This chapter is about a new *name* for life that doesn't add any new meaning to it. 'GMO' is a non-meaning of life, a term that purports to refer to something new that isn't really there.

Names matter a great deal to us. We are a very tribal species that likes to associate in groups. We band together with similar folks who share our values and beliefs. This can be a very efficient way to deal with a world overloaded with complex information. You don't need to figure out everything for yourself. You just trust others, who share your attitudes, to crowd-source it for you. Conclusions get passed around under code names. The term itself tells you whether this is a good thing or a bad thing according to your tribe. The same term will have positive connotations for one tribe, and negative ones for an opposing tribe.

Marketers know this. If you want to sell things to a given tribe, you must be careful not to use terms with a negative connotation for the tribe, but you can also benefit tremendously by using its positive terms. This tells you, the tribe member, that the marketer is one of you. Politicians do this as well. Anyone who wants to reach a mass audience with sound bites has to understand how this dynamic works. If you happen to be a member of a tribe for which 'green' and 'natural' and 'organic' are positive terms, there's a good chance that you've learned 'GMO' is a negative term. 'Green,' 'natural,' and 'organic' all have prior connotations that evoke a certain positive emotional response (for the right tribes). These inherent connotations can be tricky for companies to manage. 'Chemlawn', for instance, was once thought to be a great name for a company that treated your lawn to keep it green and the envy of neighbors. The name was conceived when 'chemistry' still had a high-tech, future technology sort of connotation. 'Chemical' and 'green' were compatible thoughts then. Over time, 'chemical' became antagonistic to 'green' and the thought of a company spraying "chemicals" on your lawn sounded like a recipe for birth defects. The company had to change its name. Recently the term 'pink slime' was just an ordinary term of art among meat processors, referring to just one of many animal-derived fillers used in consumer food products. But when the term surfaced at the consumer level, it suddenly became the kiss of death for any product containing it. Ew! Pink slime! Initially some companies protested that it is perfectly safe and has been in the food supply for some time, but they soon saw this would be a losing battle. 'GMO,' on the other hand, has no inherent, emotive connotation. It is a rather pedantic sounding, biology process word. What's that (other than a bad thing)?

It's Not Natural

The folks who oppose what they think is a scary, new form of life, that 'GMO' is supposed to designate, realized early on that this was not a particularly emotive term. And the argument behind it is even more arcane and technical. So they found more emotive symbols to convey their fears: 'FrankenFoods,' 'fish tomatoes,' pictures of fruits and vegetables with hypodermic needles stuck in them. The common core of this belief in a radically new form of life is the idea that genes aren't modified in all the other kinds of life – or put another way, that the genes of organisms weren't modified until biotech came along. This version of folk biology is shared by religious fundamentalists who don't believe in evolution. God/nature created all the original genes in a very orderly and species-separated manner. This makes genetics very natural and safe. It avoids the abomination of hybrid species, and all of the catastrophes that might result if we insist on fiddling with the natural order.

A popular website on the matter puts it this way:

> *Eating genetically engineered (GE) foods may cause defects (imbalances) within the human body. Because the molecules of hybrid and GE plants are altered, it stands to reason that the molecules of the human makeup also become altered and this altering manifests in human behavior, actions, and thoughts in*

*the form of defective or faulty behavior, actions, and thoughts.
After all, you are what you eat.*

Well, it stands to reason only if there used to be a ready supply of edible organisms whose molecules were not altered. But we've never had any of those. *Every* organism is a genetically modified organism. That's how evolution works. DNA is frequently modified by copy errors whenever a cell divides. Sometimes this changes genes. That's how we get new raw material for natural selection to operate on. Every time sexual species reproduce, the two contributing genomes are radically modified by random recombination. Again, change. These changes are mostly neutral. Sometimes they are catastrophic, leading to an unviable offspring or one that can't further reproduce. And sometimes a beneficial new feature results. We call this disorderly, random, willy-nilly modification of organisms' genes *life*. Each new organism that enters this life starts with altered genes. When you eat another organism (or another organism eats you), its recently altered genes aren't swimming in your tomato sauce. They are locked up in its DNA. Your metabolism breaks that DNA down into raw components, just like it breaks the organism's proteins down into their constituent amino acids. You might use these new parts to synthesize your own proteins and genes, but at that point they are just letters, not words or phrases.

Nature has very little respect for what we call species boundaries at any particular point in time (say now). Species are very fluid things. Nature evolves them. That's how we get them in the first place. At the bacterial level, the process is so fluid that we can't really talk meaningfully about species at all. Anyone can swap genes with anyone, and they often do. It is a more conservative affair in eukaryotes, but whole interspecies genomes get merged in single transactions all the time by symbiosis. We have already touched on many of these arrangements: the eukaryotic cell itself with its captive mitochondria and plastids; lichens which merge different biological kingdoms; plants that fix nitrogen in the soil by way of the bacteria embedded in their roots; us with our 100 trillion bacteria. There are even species of photosynthetic snails that have crossed the animal-plant divide. Nature is a very prolific genetic engineer.

OK, you might say, but this is not really relevant because humans are not involved in this kind of genetic modification. *Natural* selection is OK. It's when humans get in the loop and start intervening in the process that we get ecological catastrophes. 'GMO' really means *genetically modified* (by humans) *organism*. Fair enough. We do have a long track record of screwing things up. But humans have been engaged in genetic engineering for tens of thousands of years. This is not a recent phenomenon. The selection is not any less natural because humans are involved. We are simply part of the natural environment that selects the candidate animals and plants for further reproduction. The only real difference with pre-human evolution is that speciation happens more quickly when we are the selectors. We create strong selective pressure for the features that please us. We weren't even consciously aware we were doing this. We simply ate the naturally occurring fruits and vegetables we liked best. So did other animals before us. This caused more of their seeds to be propagated. You might even say they were selecting us. We befriended and fed wolves and wild cats that were just docile and social enough to hang around us. We unconsciously selected for their domesticated features, leading over time to fully domesticated species of dogs and cats.

OK, fine. Then what we really mean by 'GMO' is *genetically modified* (by humans consciously selecting specific new traits) *organism*. Well, we started doing this around 11-13 thousand years ago when we transitioned from hunting and gathering to herding and agriculture. We intentionally planted the seeds of the best individuals in the next season, hastening the development of the food and cultivation features that most appealed to us from the present season. We bred the animals with the best food, work, and domestication features into the next generation. Virtually every species of food crop that we have today is a human cultivar – varieties that we created by selective breeding over the last 10,000 years or so. Before we came along there was no corn, wheat, barley, rice, sugar cane, tomatoes, potatoes, jicama, yams, beans, peas, lentils, soybeans, peanuts, melons, squash, or cucumbers in the edible forms we consume today. These were derived by our ancestral genetic engineers by cultivating and crossing wild varieties that were either too small, too hard, too bitter, too poisonous, too tough, too difficult to farm, or insufficiently nutritious to be used as food. We turned them into edible foods.

The wild-type reproduction strategy of plants is to make their seeds inedible (hard, poisonous, inaccessible) and either disperse them with biomechanical mechanisms, such as wild wheat's shattering husks, or wild peas' popping pods, or embed them in edible tissue, like fruit, for dispersal through animal waste. We converted many of the seeds to edible components (grains, nuts), selected out the natural dispersion mechanisms, eliminated some of the seeds altogether from the fruit, and converted the growing seasons, and water and soil nutrient requirements to better suit domesticated cultivation. As a result, most of our current food crops would be incapable of surviving in the wild. They are now completely coevolved with human agriculture. We collectively sampled from the approximately quarter million wild flowering plant varieties, and ultimately domesticated less than a thousand of them. Today, only about a hundred of those varieties are still farmed on any scale worldwide. All of these "natural" products of nature that form the reference standard for orderly, safe gene flow before the intervention of human technology are themselves the products of human technology.

OK, but humans just *assisted* nature in the natural selection of cultivars. The transformations were still done by nature; we just guided its hand, so to speak. Humans didn't physically intervene in the reproductive materials or mechanisms. So 'GMO' really means *genetically modified* (by humans consciously selecting new traits by directly intervening in the process of creating new genes) *organism*. But we did that too. Breeders took pollen from one variety of plant and used it to fertilize different varieties of other plants. We took cuttings, and surgically grafted two plant varieties together to form a whole new combined genome – something that would never happen naturally.

OK, but we never mixed different *species* until now. That doesn't happen in nature. How about *genetically modified* (by humans consciously selecting new traits by directly intervening in the process of creating new genes, including the mixing of species) *organism*? The "mixing species" thing seems to be a real hot button. Second only to the pictures of crops with hypodermic needles in them, are the photoshopped images of two different fruits combined in a montage with half the characteristics of each – presumably what a FrankenFruit would look like at the phenotypic level. But you don't need to resort to photo editing. There are plenty of very colorful photos of real FrankenFruits and FrankenVeggies that humans have made prior to biotech, among them: the tangelo (tangerine and grapefruit), the blood lime (lime and orange), the tayberry (red raspberry and blackberry), the pluot (plum and apricot), the pecotum (peach, apricot, plum), the rangpur (orange and lemon), the limequat (lime and kumquat), the rabbage (top half radish, bottom cabbage), the grapefruit itself (pummelo and orange), the pomato (tomato and potato), the Fuji apple (apple and grape), and broccolini (broccoli and kale).

Except for the rabbage, these hybrids look rather elegantly integrated. Frankenstein integration suggests awkwardly stitched together components to make a very unnaturally looking composite life form. But there are plenty of these too. One that I am familiar with, having lived in California, is a cross between the black walnut tree and the English walnut tree. I lived in a valley that had once been a walnut orchard. Because California is semi-arid, there was not enough water to sustain the English walnut trees that produce the big nuts we are familiar with. The black walnut tree, however, has a much more drought-resistant trunk and root system, though not any kind of nut that you would want to eat. So growers discovered they could graft the bottom of a black walnut to the top of an English walnut and get the best genes from both. But the result does look rather Frankensteinian. The black trunks suddenly stop at a very irregular splice point before the grey branches of the English begin. Oh, did I mention there was no biotech involved?

But it wasn't done in a lab. *That's* the difference! We're just opposed to *genetically modified* (by humans consciously selecting new traits by directly intervening in the process of creating new genes, even by mixing the genes of different species, but doing it in a lab) *organisms*. Shall we go on? There's something they do in labs apparently that makes all the difference. Something about the process of modifying genes in a lab conveys "lab cooties" to the new organism – a colorless, odorless something or other that you can't detect by inspecting the outcome. But whatever it is, it's certainly not natural, so it's bound to cause problems.

This is the fundamental problem that regulatory agencies would have, even if they wanted to quarantine this new variety of life for special treatment. "Made in a lab" doesn't tell you anything of interest about what it is that you've made. Opponents of GMOs struggle to fill in this story about the special, universal contamination of lab procedures, but it's hard if you are not familiar with the science or the techniques (or with the way nature does it at the molecular level). A favorite account raises the specter of the "gene gun" that researchers use to "shoot" genes into an organism. That can't be natural! There actually is such a device, used in some GMO procedures, but don't worry. It leaves no gunpowder (perhaps, biopowder?) residues in the organism, or any gaping wounds. Finding the common safety defect in *all* lab procedures for genetic engineering faces the daunting array of hundreds of procedures for re-engineering everything from bacteria that produce biofuels, to putting fluorescent genes in model organisms to observe cell dynamics, to repairing genetic abnormalities that cause human diseases. Each year that goes by adds probably several dozen more procedures to the pile.

So this new class of life, that contains no biological properties that are not already in the previous class of life, and this new, heterogeneous category of manufacturing processes that uniquely create this redundant form of life, don't add much to the safety debate. We still face the problems that come with introducing novel things into our ecologies, living or not, artificially or naturally produced. What matters is *what* we are introducing. What are its properties? The presence of a GMO label on the package tells you nothing about this. You don't know what you're actually getting, whether what you're getting is a good thing or a bad thing, and whether or not you also get it in non-GMO foods. The "contains no GMOs" label is conversely vacuous, except to alert you to the fact that it is kosher (or not) with your tribe.

The Curious Case of Bacillus Thuringiensis

If you are really concerned about this issue, independent of what your tribe says, you need to look at what feature is being introduced into what food, then find out something about the health effects of that feature. Ironically, most of the GMO crops that have made their way into our labeled food so far weren't engineered to add a new food feature (those are added in the old-fashioned way by the more traditional food processing industry). GMO crops were re-engineered to make the original crop more resistant to specific pests during cultivation. So the food safety issue is whether the new gene that expresses the pesticide when the crop is grown somehow leaves a residue in the final food, and whether that is any sort of a health concern for humans. The first battles were fought over the introduction of such a gene into corn. This became known generically as 'bt corn,' derived from the name of the bacterium that contributed the gene, that produced the protein, that killed the insect, that ate the corn, that lay in the house that Monsanto built. Variants of the same gene have been approved for use in 'bt potatoes', 'bt cotton', and 'bt soybeans'.

'Bt' is an abbreviation for *Bacillus thuringiensis*, a bacterium naturally living in the soil all over the world. As we have already noted, bacteria have been around for a long time, and their constant competition for survival with other micro and macro organisms has caused them to develop a wide variety of molecules that are toxic to their specific competitors. Scientists see this as a great source of already effective antibiotics, as well as an ongoing, large-scale experiment in the vetting of the effects of these antibiotics in ecosystems. I am personally acquainted with B. thuringiensis because I recruit their help each July to keep the flowers on my geraniums and petunias from being eaten down to the stems in a matter of days by caterpillar larvae. When the annual attack comes, I never see the moths, or their larvae, just the devastation of the flowers. It took me a few years to figure this out, but only a few hits on the Internet to find the natural solution: Bt. Gardeners and organic farmers hail it as the ideal pest control not just because it is an existing, natural solution, but because it is so specific to particular species of insects. It doesn't affect other, beneficial insects, or vertebrates.

You can buy it at your local garden supply store in the form of a viscous, brown liquid that you simply mix with water and spray on your plants. The brown liquid is a concentration of the Bt bacteria in their endospore form. They sporulate under conditions of environmental stress, such as lack of food, and will switch back to active mode when conditions are right, so they can just sit in the bottle for years. Their caterpillar toxin is a crystal protein, named Cry, that is expressed by a gene and remains inside the bacterium.

When caterpillar larvae eat the bacterium, the acid level of their guts breaks down the crystal and exposes subcomponents that will bind to specific targets in their intestinal walls, causing perforations. You have to have the midgut pH level of an insect and the specific molecular binding site in your intestinal walls to be affected by it. So it's tailored to a very specific adversary. I have no animosity toward this particular species of caterpillar in general, and they, no doubt, know nothing of me. It's just that we are involved in a zero-sum game concerning my flowers. So rather than dropping bombs, or sending in troops, I just try to nudge an existing, natural conflict a little more in a direction that aids my interests. I provide a little logistical support for one side's troops, helping them get into position.

This is what makes Bt the universal insecticide of the organic farming industry. It may surprise you, if you are in the organic tribe, that organic farmers use fertilizers and pesticides. You may have thought that the 'organic' label meant "grown with none of that stuff." But it really means, "grown without *synthetic* versions of that stuff." Natural versions are OK. (Actually, you are allowed to use the 'organic' label even if you use the synthetic stuff as long as you stop for a prescribed period before harvest). So given the prior 40 year track record of the Bt Cry protein in agriculture, both for its benign environmental impact and human health characteristics, this seemed like just the kind of thing you might want to re-engineer into corn (and tobacco, tomatoes, potatoes, cotton, and soybeans). By borrowing the Cry gene from B. thuringiensis, corn cells would express the Cry protein endogenously in their tissues. The one key difference in the delivery method of the protein to consumers, pests and us, is that in the GMO version you don't need the intervening bacterium. In the spray method, any residue on the food that gets all the way to your mouth will include the whole bacterium. This is no more of a problem for humans (we think) than the protein itself, but the GMO delivery method does rule out the very unlikely scenario of the ingested bacterium making it all the way to your own gut microbiota where it's conceptually possible to swap genes.

So you may be wondering, what's the net difference in your exposure to Bt (except for bacterium+protein vs. just protein)? Do you get more or less of the protein from the organic method vs. the GMO method? Well, we don't know, and it looks as though we are not likely to find out. The USDA doesn't traditionally monitor pesticide levels in organically produced foods, only conventional ones. They are about to start, but only to monitor for levels of conventional pesticides to encourage compliance with existing organic rules. They apparently do not have sufficient resources for separately detecting organic residues such as Bt. The background level of Bt already present from organic uses is not well known. Assays by researchers in Denmark have found Bt on fresh fruits and vegetables for sale in retail shops. Assays from China have found Bt in pasteurized milk, ice cream with peach pulp and juice, and in green-tea beverages. A Canadian assay found a significant increase of Bt in human nasal passages after Bt spraying, even outside of the spray zone. So we know it's out there, in our food, and in us. Until attention got diverted to its use in GMOs, this was apparently just fine.

In 2011, a study in Quebec made headlines about the GMO-Bt issue for all the wrong reasons. It purported to detect the presence of the Bt corn protein in the blood of women, with greater concentrations in pregnant women. Oh no! Besides being criticized for using an assay method that would not be sensitive enough to detect the protein, the authors offered no evidence that the source of the protein was GMO corn. They had just assumed this. Given the background rate of Bt from organic farming, this is a far more likely source. Without the surrounding milieu of the GMO controversy, this would have been a non-result, showing the presence of an organic-friendly protein already deemed to be safe. But tribal warfare is often insensitive to such details. Although the paper has since been discounted by the scientific community, GMO opponents still cite it on their websites as if it were fresh, missing the implications for internecine warfare. If this protein were actually toxic to humans, and really was found in blood, this would be a safety concern for *organic* foods. Conventional pesticides don't have this exposure. The same tribal confusion was reported by the Organic Consumers Association in a 2006 article warning that "Bt crops threaten public health." The basis for their claim was a 1991 study focused on human exposure via inhalation of Bt sprays. The results showed immune responses and skin sensitization to Bt in 2 of 123 farm workers. Oh no! GMOs cause allergies! They somehow missed the point about the vector being *sprays* of Bt – an organic farming side effect.

Tribal passions do not make for good science. And here we have a case where they actually take you in the wrong direction. You'd like to know if Bt proteins are safe for humans. The science says so, and they apparently were fine for 40 years or so. For now, that's about as deep as you can get, but the 'organic' and 'GMO' labels are telling you approximately the same thing on this issue.

Some Real Concerns

Fear of the unknown is a debilitating fear. It can't be assuaged. We are naturally suspicious of change and of what we can't control, but if you can't quantify that change, there is nothing you can do about it. The fear of GMOs based on their metaphysical properties, such as their unholy violation of the natural genetic order, can't be fixed because you can't possibly measure these things. You are susceptible to manipulation by cartoons of genes swimming in your tomato sauce. Genes have been swimming in your tomato sauce for as long as there has been tomato sauce. Organisms mostly eat other organisms. That's how ecologies of life work. It is only the bacteria, at the bottom of the food chain, that eat non-organic things, like sulfur and iron. The rest of us eat each other, so we are constantly eating each other's DNA. We know how that process works. It is not something new. What is relatively new in our food sources are the additives used by the processed food industry. There is nothing wrong with them per se, it's just that we don't have any prior history of passing these things through our ecologies. Compounds such as polysorbate 80 are new to the food chain, and insecticides such as DDT are new to the environment. Genes, and their proteins, from other species are not new to the food chain, and Bt is not new to the environment.

To protect our health and our ecosystems from unintended consequences when we introduce a new organism, we need to focus on what is genuinely new, and the rate of change in the balance of things that are old. GMOs, as a category of life, are not very helpful here. Their only common feature – that they involve changed genes – doesn't distinguish them, or their rate of novelty, from non-GMOs. And the one feature that distinguishes them from non-GMOs, their biotech manufacturing origin, is too hopelessly heterogeneous to imply anything about their downstream effects. If anything, GMOs, as a class, are likely to introduce less genetic novelty than traditional breeding techniques because the process is both more specific and more precise. We have a better understanding from the beginning of just which genes to change to achieve the desired phenotypic feature and what ancillary genetic and protein changes might be involved. Traditional breeders don't know this. They have to take a shotgun approach to the genome of the new organism because they can only deal in the phenotypes. If GMOs were to replace traditional breeding, it would probably result in a net decrease in the rate of novel gene combinations, because GMOs change fewer genes. On the other hand, GMOs are likely to increase the rate of whole novel organisms (with fewer gene changes), because it takes less time, and the explosive growth of biotech puts many more laboratories at our disposal.

A lot of the tribal warfare surrounding GMOs focuses on legitimate health and environmental concerns that are orthogonal to GMOs themselves. Bt corn, for instance, was claimed to kill Monarch butterflies because the Cry gene in its pollen could theoretically reach these insects in nearby fields. Monarchs were indeed on the decline, but it was not due to Bt pollen from GMO corn. This and many other aspects of negative ecological impact were raised and refuted by careful studies, but unfortunately the focus has been on who was telling the truth, who was using pseudoscience, whose research may or may not have been influenced by partisan motivations. These are also problems for any public issue needing objective scientific vetting, but a crucial point that gets lost in the noise of battle is that *even if* a particular variety of bt corn *had* been the cause of one or more of these negative impacts, this has nothing to do with GMOs themselves. It would have been about the introduction of a new corn variety with an endogenous pesticide that caused unintended ecological side effects. Naturally occurring plants express endotoxins that they have evolved to compete with ambient organisms. Food crops have been bred by traditional means to express endotoxins for specific pests. The source of the endotoxin doesn't matter. What matters is the introduction of the endotoxin into a new habitat. If anything, GMO origin of the endotoxin would be a net positive because the specific phenotypic difference in the new crop variety will be known in advance.

Similarly, generic concerns have been raised about the potential for GMOs to promote monocultures. These arise when the previous genetic diversity of an ecosystem becomes dominated by one or two varieties. The population is then less able to survive changes to the habitat, even risking extinction. A well-known example of this is the Irish Potato Famine in the mid-19th century. Two monocultures were the major cause of mass starvation among the Irish peasantry, which ultimately reduced the Irish population by about 25%. The potato had become the single dominant food source for much of the population, and of that, only two varieties of potato. A single fungus affecting both varieties swept through Europe and devastated the potato crops. Other European countries were able to tolerate it because of greater genetic variety in their potato crops, and because of a greater variety of food crops in general. Except for the occasional, natural domination of a gene pool by a dominant or invasive species, monocultures are usually brought about by humans. It is a common practice of large agribusinesses because of economies of scale. The issue tends to tar GMOs in particular because Monsanto is a dominant player in the GMO business. But again, this is an issue orthogonal to GMOs. The issue is with monocultures and whatever trait might make a crop dominant or invasive, or whatever market conditions might make one company's seeds a virtual monopoly. How the dominant trait gets into the crop in the first place is irrelevant.

Yet another orthogonal concern for GMOs is biological patents, and the conditions certain companies place on farmers to create dependencies and to dominate markets. And again the issue traces primarily to the involvement of Monsanto. These are questions of intellectual property laws and markets. I personally disapprove of biological patents, but remember, this is a useless book. I have no tribal axe to grind. I also disapprove of software patents. And these also have nothing to do with GMOs. It is hard to keep focused on the ball in this very multi-issued landscape, but stay focused we must if we genuinely care about the outcomes. Fix the things that need fixing. Look for the characteristics that matter. Find out how your tribe comes up with its prescriptions. These GMOs are not the droids you're looking for.

A label on your food that tells you it contains some ingredients that were made on Tuesday wouldn't be very useful. The GMO label conveys a similar amount of information.

13 | Love

Seize the moments of happiness, love and be loved! That is the only reality in the world, all else is folly. It is the one thing we are interested in here.
    – Leo Tolstoy, *War and Peace*

Love is just a four-letter word.
    – Bob Dylan

Perhaps our most endearing manifestation of the human soul is the soul mate. We are poignantly aware of the role of biology in this relationship, one we share with sexually reproducing life forms in general, but we like to believe that humans add a unique, non-biological dimension to this affair. What other species writes sonnets, launches a thousand ships, abdicates a kingdom, waits a lonely lifetime, counts the ways, parts with such sweet sorrow? We have raised romance to a high art, an ultimate happiness, an end in itself rather than a need to reproduce. Our higher notion of love bleeds over into non-sexual forms such as friendship and parent-child relationships. Not quite like them, but not completely different either. It is a complex emotion that defies even sociological analysis, so it seems as though this would be one of the more difficult features of the soul to account for in terms of molecules and hormones.

We haven't had much insight into the molecules and hormones of love until fairly recently, so this is not a gestalt shift with which we have much familiarity. We have, however, confronted a more accessible biological gestalt shift for as long as we have written accounts of human behavior: the distinction between love and lust. Our literature, our religions, and our social customs are thoroughly imbued with this bipolarity. We celebrate both in varying degrees (though lust quite less often in religion), but the lusty aspects have an uneasy identification with animal behavior. One glance down the animal kingdom reveals that mating is not a very romantic affair. It is characterized by violence, coercion, infanticide, promiscuity, power, and capitulation. There is no modesty, no privacy, no endearment, no lingering afterglow. It is a brief, very utilitarian exercise. It perpetuates the species. It works very well. But we are more refined, we believe. All of those other species have mates, but only we have soul mates. Long before we knew about hormones, we have been insisting that love is more than just animal lust.

As with empathy in the next chapter, our higher notion of love seems to be a kind of emotion that can't possibly evolve via natural selection. If natural selection favors those that promote the survival of their own genes, how do we account for behavior that sacrifices one's own interest in favor of another? The evolution of sexual desire is easy to understand. The more you do it, the more you reproduce your genes. But this higher form of love is only coincidentally connected to reproduction. That is not essential. The emotion often does not encompass this at all. The emotion can support behavior that runs counter to reproduction (e.g. contraception). It is an affection that promotes disregard for one's own interest, a willingness to promote the lover's interest. What possible survival value was added by tacking this motivation onto the already well-oiled mechanism of sexual desire? It seems like it should convey negative fitness on those that develop it.

Well, humans are indeed outliers in the animal kingdom. We have evolved interesting features found nowhere else (but you knew that). There is an evolutionary cause for how we got this way, though as with much else in our unique emotional makeup, it is not one cause to one emotion. Our emotional nature is a huge composite of competing emotions that we got by successively layering new emotions over old ones. Some of these competing features are specifically biological, or neurological, but we also have a dominating layer of social emotions that are informed by language and culture. We evolved the biological *capacity* for language and culture, but which language and which culture is something that gets imprinted on us by our environments. So the variations are unbounded. Mix this all together and you get our complicated, often contradictory attitudes about love and sex (and almost everything else).

The Mating Game

Everyone understands the need for sex. Even the most prudish of religions buy into this aspect of evolution. Sexual desire and pleasure have often been regarded as sins in themselves, but the acts they support are clearly necessary to be fruitful and multiply. A religion can promote celibacy as a virtue only for certain designated subsets of its members if it expects to survive long term. Sects that didn't buy into this basic biological fact, the early New England Shakers for instance, are understandably no longer with us.

We often characterize natural selection as survival of the fittest, but survival is only an aid to the selection process. Reproduction is what matters. It is really survival of the most fruitful. If you survive, but don't reproduce, you don't matter. If you fail to survive, but manage to reproduce first, you've made your mark. Strength, size, fighting prowess, adaptability, and other characteristics that keep you alive longer than your competitors only matter insofar as they support your reproduction. Differential reproduction is what determines the winning genes. If you can reproduce faster or more often than your competitors, the other traits don't matter. This is why bacteria are so good at the evolutionary game. And sex is not even involved in their reproduction.

Sex is recombination, not reproduction. It represents a reduction in the number of outstanding organisms because two of them become one. Recombination creates novelty much faster than reproduction, which is why it evolved in the first place. Reproduction can create novelty only through occasional small mistakes. But you can combine without reproducing, and you can reproduce without first combining. Every possible arrangement of these two processes is found somewhere in the various biological taxa: asexual reproduction, merging without dividing, individuals of separate genders combining then reproducing, individuals with separate-gendered parts combining them to reproduce, individuals doubling their own genome then reproducing, individuals that reproduce both sexually and asexually. Large animals tend to do it only one way, by combining cells from two individuals of opposite genders and then reproducing the combined cell trillions of times to make a single individual. But even here there are exceptions. There are documented cases of sharks giving birth by parthenogenesis, where the female creates a zygote by doubling the genome in one of her germline cells. Now that we have perfected the technology of in vitro fertilization, recombinant DNA, cloning, and iPS cells, the "natural" method of animals has become entirely optional. We tend to think of our particular pattern of gender, or even that there are genders, as a fundamental feature of life, where it is in fact just one particular variation in a much larger universe of reproduction strategies.

But if your species depends on sexual reproduction, as ours does, then it's a good idea to have a native incentive to do this. It takes some effort, and it entails some costs, so it's no accident that the genders of sexual species have a strong, inbuilt attraction for each other, and a strong desire to find and mate with partners. Since sexual reproduction goes all the way down the vertebrate line (and beyond), much of the basic neuroanatomy implementing sexual desire and behavior is in the more primitive brainstem portion of our brains that we share with distant ancestors. We don't need to learn what to do in sex; it is already encoded. If anything, we have to learn, via human culture, how to selectively ignore or control these behaviors. Where, when and how to use these basic mating programs is controlled by three separate, but related neural motivation mechanisms: the motivation to seek and mate with partners of any kind; the motivation to select a certain preferred partner to mate with; and the motivation to hang around (or not) after mating. Evolution can tune these three motivations to give different species different mating systems. A mating system determines the way in which a social group is structured in relation to sexual behavior. Variations include *monogamy* (one male, one female in an exclusive relationship); *polygamy*, which has three varieties, *polygyny* (one male in an exclusive relationship with multiple females), *polyandry* (one female in an exclusive relationship with multiple males), and *polygynandry* (multiple males and females in an exclusive relationship); and *promiscuity* (anyone mates with anyone).

Now, depending on how your particular species handles the mechanics of fertilization and development, males and females may have similar costs and benefits to reproducing their genes, or they may have radically different ones. If you are a fish, for instance, where the genders separately release sperm and eggs so that fertilization and development takes place in the external environment, the relative investment between the genders is about the same. They both expend about the same amount of time and resources, and they have an equal interest in the survival of the offspring. If you are a bird, the female spends a little more on the internal gestation of an egg, but once it is laid, the genders are again on pretty equal footing. Either can incubate the egg(s), and either can feed the fledglings. If you are a mammal, however, the female pays an enormously disproportionate cost. She must gestate the entire development of the offspring inside her body, and is the sole source of early food before weaning (that's why we're called mammals). There's not much a male can do, or needs to do, other than protecting the female during this phase. And in the interest of promoting his genes, he actually has an incentive to keep mating with other females during this period. The female has no choice but to gestate and wean the offspring because she can't reproduce during this period, so her genes do not survive unless the offspring survives. The male can reliably bank on this and get a free ride for his genes while pursuing a diversification strategy of setting up more free rides.

As you can imagine, then, the reproductive "strategies" of male and female mammals are very asymmetric. Biologists often account for how evolution produces a given mating system by viewing a given species' situation as an evolutionary game. Given the starting conditions, and respective mechanics of reproduction, the genders will evolve behaviors that correspond to the best strategy they *could* adopt given the best strategy of the other. This includes your strategy toward opposite-gendered partners, and your strategy toward same-gendered competitors. The males and females of a given species are playing a game with each other to see who can get the most of their genes into the next generation. It may seem odd to think of animals playing such a game with explicit strategies in mind. They aren't of course. Animals, with the exception of humans, cannot think ahead to the reproductive consequences of sex. They are motivated by the emotions that attend the behavior itself (as are we, largely). But genetic variations that tweak their sexual motivations, in the three dimensions mentioned above, will cause them to behave in certain predictable ways relative to each other. Evolution will preserve those motivation/behavior pairs that result in stable winners of the "preserve my genes" game. So whether organisms know it or not, they are playing a gene preservation game. And quite unbeknownst to them, their heritable behaviors correspond to strategies in that game. Evolution will produce the same result, without any foresight, that you would get if the participants were actually aware of the game and consciously pursued the best strategies, because the winning genes will encode those strategies.

Placental mammals, like us, have been playing the mating game for about 65 million years. The surviving species today exhibit a broad commonality in their mating systems. Monogamy is extremely rare. About 97% of species do not form lasting pair bonds. Promiscuity is the norm for both males and females. Mating is typically performed in the open, a public event visible to the entire social group. Females take on all of the responsibility for rearing the young, either singly or in groups, and males often leave the social group after mating. Even those who stay do not typically recognize their own offspring. Mating behavior for both males and females occurs almost exclusively during the portion of the female ovulatory cycle in which pregnancy is possible, or most probable. In some species, it is not even physically possible to mate except when conception is possible, but even in those species that can, neither gender shows much interest in mating outside the cycle. Females often publicly advertise their fertile phase with swellings, odors, sounds, and behavioral postures. These clues, in turn, start the males in motion. In monkeys, for instance, a surge in female estrogen at ovulation time will stimulate the cells lining the vagina to secrete molecules that result in a change to the population of vaginal bacterial species. These bacteria, in turn, secrete the magic odors that male monkeys find irresistible. Post ovulation, when the microbiota population returns to normal, the odors repel the male. Another little bit of essential coevolution with our bacterial symbionts.

Human mammals, on the other hand, are the extreme outliers in mating systems. We are mostly (serially) monogamous. Our mating is intensely private. Couples form long-lasting relationships. Men hang around to help rear their children, and often hang around even in the absence of children. Neither men nor women are aware of ovulation, so the condition cannot be advertised (this is called *concealed ovulation*). Couples engage in sex at all points of the ovulatory cycle, regardless of the possibility for conception. We also engage in sex when conception is strictly impossible, such as during pregnancy and after menopause. Menopause itself appears to be a uniquely human trait in nature. Female sexual productivity declines with age in other species, but never to the point of outright cessation. Each of these outlying traits, except for privacy, can be found individually, here and there, in other mammal species. Prairie voles and gibbons are monogamous, orangutans and vervet monkeys have concealed ovulation, and bonobos and dolphins engage in recreational sex. But no one puts it all together like us. So why did we play the mating game so differently?

Molecules and Hormones

Well, our most characteristic feature as a species, our outsized brain, accounts for a lot of the difference. It has nothing to do with being more intelligent; it is simply a matter of size. Our development of the big brain had some associated costs and constraints. Put simply, the amount of brain needed to support human intelligence just doesn't fit inside of a fetal skull. As our brains continued to get bigger, the human female anatomy evolved successively wider hips to accommodate a bigger head coming down the birth canal, but eventually reached feasibility limits for the mother. The significantly higher rate of knee injuries in women's sports attests to the already non-optimal angle that these wider hips direct toward the knee joints. So to accommodate the growing brain size, human infants are born approximately three months premature relative to their chimpanzee counterparts. The brain has to continue growth and development outside the mother. This makes human infants absolutely helpless in the first three months of life. The continued development and expansion of the brain up through age five makes them slow to develop the fine motor control for efficient walking and finding their own food (or evading predators in the environments in which we evolved). The progeny of other primates, and even more so in the lower mammals, hit the ground running so to speak. They can either survive on their own, or with minimal or short-term maternal assistance. Human infants were pretty much toast without the active help of both parents. So males were constrained in their philandering on two counts. Because of concealed ovulation, you could never be sure whether you had successfully planted your genes, so you couldn't reliably double dip. You also couldn't be sure someone else hadn't planted his genes in your absence. And if you didn't stay around to help rear your offspring, there was a good chance your female partner couldn't do it alone, so you would both lose the gene game.

Although pair bonding is rare among mammals, it is not as simple as having highly needy offspring. There is some genetic evidence that humans were largely polygynous until 10-20 thousand years ago, though this is well within the range of influence by social culture. Prairie voles have the typical quick-to-develop offspring, and no concealed ovulation, so males seem to have more fruitful strategies available, but they form life-long pair bonds instead. They have related cousin vole species without pair bonding. We don't really know how they evolved this particular strategy (or even exactly how we did), but because they have, prairie voles have become the model organism for studying pair bonding in the laboratory. Because of them, we have a decent understanding of the neural correlates and hormones that underlie it, aspects of biology that we can apply to ourselves.

Monogamy has evolved independently 61 times in mammals, and is very common in birds, because you can get it by toggling a few well-defined switches in neuroanatomy. Evolution doesn't have to completely reinvent what might seem like a counter-intuitive strategy. It can repurpose some neural programs that have overlapping characteristics by making relatively small changes to the genes that express certain hormones. It appears that we have three overlapping neural systems that bear on reproductive behavior: one to control mating interest, one that can regulate mate selection, and one that can create pair bonds. The first of these is sex specific, but the other two are intertwined with other cognitive and motivational systems. By mixing and matching these systems, you can get a large variety of strategies for the mating game. It also turns out that the distinctions and interactions between these three neurobiological systems match our traditional distinctions and interactions between love and lust.

The sex specific system is one that all vertebrates have. It is governed primarily by *testosterone* in males and *estrogen* in females. Surges in the release of these hormones are what drive sexual desire and coordinate the inbuilt behaviors for mating. In mammals that have evolved to mate only at the peak of ovulation, it is the female menstrual cycle that drives the elevated release of estrogen. This causes physiological changes that make the female receptive, and that produce the positive signals for males. These signals, in turn, start the male's testosterone flowing and the dance begins. If promiscuity is the rule, and neither gender expresses a differential preference for mates, this is enough.

But many mammals, and all birds, express preferences for particular mates. Biologists call this courtship behavior. We are most familiar with this (in non humans) in the elaborate courtship rituals and displays of birds. Not surprisingly, most bird species form lasting pair bonds. Birds are special in the mating game both because males and females have a roughly equal investment in the offspring, and because their avian lifestyle makes them significantly less exposed to predators, and extremely mobile in the search for food. They can afford to evolve sexual preferences, such as the peacock's tail, that don't relate directly to survival value. Their elaborate colors, and songs, and dances, and ornamentation evolve in a feedback loop where a female preference for a particular display characteristic will select for males that have more of it. The progeny reproduce both the preference (in females) and the display (in males). This is where survival of the sexiest can outrun survival of the fittest. In species with riskier life expectancies, the displays of males are usually proxies for some fitness characteristic, so the female is implicitly selecting "good genes." But "good genes" is relative to the environment in which they propagate. The only universal is "genes that will get reproduced."

Courtship behavior is hard to miss in birds (and in humans), but it can be more subtle in non-human mammals. This is how one research paper describes it. "Courtship attraction is characterized in mammals by increased energy, focused attention, obsessive following, affiliative gestures, possessive mate guarding and motivation to win a preferred mating partner." The neural circuits that underlie this motivation system are regulated primarily by *dopamine* and *norepinephrine*. The perception of a possible mate with just the right characteristics sets off a surge in dopamine, the same neurotransmitter and the same reward centers, incidentally, that figure in drug addiction (Robert Palmer was right). What it is about this potential mate that makes it so attractive can be a function of innate preference, individual variation in preference, and learned preferences, but it is dopamine that signals a hit. When a female prairie vole finds a mate, for instance, she experiences a 50% increase in dopamine. If you artificially interfere by injecting a drug that blocks her dopamine receptors, she will no longer prefer the mate. If you inject a drug that activates her dopamine receptors, on the other hand, she will begin to prefer whatever male happens to be present at the time of injection.

Kind of takes all the romance out of it? Well, the romance is still there in the affection for a particular partner that dopamine promotes with its generalized reward/addiction system. This affection is normally connected to specific traits of the partner that she finds attractive. We have simply discovered the switch that connects these two things. Evolution can experiment with different varieties and manifestations of love by fiddling with this switch, connecting the affection to different kinds of traits. In humans, and in mammals all the way down the line, this romantic affection system is linked with the sexual arousal/behavior system in both directions. Increased dopamine often causes elevated testosterone and estrogen, and increases in these hormones cause elevated levels of dopamine and norepinephrine. Just what you would expect. Love and lust are intertwined. They are both capable of occurring independently, in the right circumstances, but in general, each promotes the other. If I mate with you, I may become addicted to you. If I become addicted to you, I am likely to want to mate with you.

The third neural system of love, the one that controls pair bonding, is primarily regulated by *oxytocin* and *vasopressin*. Oxytocin has been variously called the "love hormone," the "cuddling hormone," and the "bonding hormone." Its general effects are a lowering of interpersonal social inhibition, an increase in interpersonal trust, and the establishment of a lasting, affective bond between two individuals. This extends to parent-child relationships, and to relationships with same-gendered friends. It is released by intimate contact, following orgasm in humans, by skin touching, kissing, cuddling, and grooming (in non-humans). Like dopamine in the romantic context, it causes a sort of imprinting on the specific partner present at its release. Engaging in the close contact behaviors that cause its release will dispose the participants to bond with each other. This can also be cruelly exploited in the lab by artificially interfering with the oxytocin and vasopressin receptors to cause imprinting on the nearest individual, even though that individual didn't come through the customary channels.

As with the other two systems, this one can act independently of romantic or sexual attraction, as when it forms filial bonds with same-gendered friends, but it also has connections to the other two. Dopamine stimulates oxytocin release, and oxytocin stimulates dopamine release. So the behaviors associated with each tend to promote the other. But it is a complex relationship. Each can, in some contexts, interfere with the expression of the other. The gene in prairie voles that expresses vasopressin receptors for pair bonding has a certain degree of variation, allowing for different individuals to express varying densities of the receptors. This density corresponds to greater or lesser degrees of strength of the pair bond. Humans have a similar individual variation in this gene, which makes us more or less prone to pair bonding on an individual basis.

Experimentation can be a lot more invasive with prairie voles than with humans. So humans' neural correlates of love are studied with functional magnetic resonance imaging (fMRI). This is essentially an MRI for the brain that shows where the dominant blood flow occurs when a subject is observing a certain stimulus or performing a certain task (that doesn't require head movement). A number of studies have been done using male and female subjects who are "deeply in love," some for relatively short periods and some for as long as two years. Their fMRIs are taken while viewing images of their partner, as well as images of same and different gendered friends or strangers. Since any given emotional response will light up most areas of the brain, the differential effects of love are measured by subtracting (via computer) the areas lit up by non-lovers from lovers. This (theoretically) leaves the areas that are specific to the love response. These as well as differential fMRIs of subjects undergoing sexual arousal feature the same regions that support testosterone/estrogen, dopamine/norepinephrine, and oxytocin/vasopressin responses in laboratory animals. The arousal pathways are distinct from the love pathways, and the "new love" vs. longer-term pair bonding pathways are related but become distinct with differences in the length of the relationship.

Another study, attempting to approximate the drug-induced, fiddling-with-the-switches experiments on prairie voles, used fMRI scans of human males who reported being in a relationship with a lover for at least 6 months. They were imaged viewing photos of their lover, and various other women whom they knew and whom they did not know. The photos were chosen by independent reviewers who rated all of the women as being equally attractive. One group was given a dose of oxytocin via nasal spray, and the other, control group a placebo spray. In the males under oxytocin influence, the dopamine oriented pleasure and desire centers of the brain lit up strongly for the lover, but not for the equally attractive strangers, suggesting that oxytocin increases the attractiveness of an existing partner. In prairie voles, the oxytocin intervention also makes males slightly aggressive toward non-bonded females.

All of this dovetails nicely with what we are already familiar with in human courtship. We recognize the distinction (and the interrelationship) between love and lust, but we also have a sense of the distinction between the excitement of a new love and the slower burning attachment of a long-term soul mate. We are painfully aware of how few of the former transition to the latter. We have individual variations in our capacity for each of these three aspects of love, but it appears we are wired so that each tries to promote the others. The features that attract you to any member of the opposite gender are something your whole species agrees on. The features that first attract you to a particular partner are in some sense objective, but a little subjective, reflecting your individual differences. Then the relationship itself can convert the attractiveness of your partner into a very subjective perception that makes your mutual attraction effectively unique.

From the hormonal point of view, the sexual drive that we share with all animals is indeed sex specific, but the other two systems of romantic attraction, what we like to think of as our higher form of love, are borrowed from more general emotional systems of euphoria/addiction, and inter-personal bonding. We recognize this crossover in our romantic behavior, even without reference to the hormones. Stir them all together and you get our familiar, contradiction-laced concoction called love. But this also makes it likely that we share this broader range of emotions with other creatures like birds and prairie voles. It is not just a human concept. If they had language, perhaps they too would write sonnets.

Your Inner Mammal

Given our deep mammalian heritage, it is not surprising that we retain vestiges of mating preferences that make more sense in the context in which we evolved than in our present context. We have become accustomed to social and political equality between the genders, or at least the goal of such equality. We reason that men and women should have similar aspirations and similar goals, and that there are no inherently asymmetrical roles between the genders. We resist the cultural asymmetries of our recent past that assign gender-specific roles and expectations, believing these are inessential because they are learned. Yet we still often behave asymmetrically in our mating preferences. Men pursue and propose, women entertain offers and accept or reject. Men are more easily interested in sex, women more in establishing relationships. Where age differences are involved, men prefer younger women, and women prefer older men (generally).

Each of these asymmetries makes sense in the context of how our mammalian ancestors played the mating game. Millions of years of males spreading their genes around, and females seeking help with offspring, leaves a trail in our mating motivations. A 1989 survey, for instance, measured this interesting gender difference among students on college campuses concerning the linkage of the attraction response and the mating response. Attractive men and women were hired to approach students of the opposite gender with this opening line: "I have been noticing you around campus. I find you very attractive." The actor then followed up with one of three proposals: "Would you go out with me tonight?" "Would you come over to my apartment tonight?" "Would you go to bed with me tonight?" About half of both men and women responded positively to the date proposal. But to the "come over tonight" proposal, only 6% of women agreed, where 69% of men did. And on the "sleep with me" offer, all women declined, but 75% of the men accepted.

The features that attract us to each other would seem, on the face of it, to be very culture dependent, century dependent, and age dependent. Attractive fashions, dress, and hairstyles come and go in trends. Who displays the peacock's tail has changed from men to women over the centuries. What physical features of face and body are attractive varies with race, generation, and ethnic background.

But there are human universals. It seems that we all prefer symmetrical faces. Virtually all cross-cultural surveys that ask participants to rate photos of faces for attractiveness converge on this result. Another, related phenomenon is that composite/averaged images of actual faces are generally preferred to the individual images. The more photos are added to the composite, the more attractive the result. This is possibly the same preference for symmetry because the composite averaging reduces individual variations. Theorists struggle to relate this to some fitness advantage of symmetrical individuals that we may have evolved a preference for. Because we are bilaterally symmetrical animals, there probably is a correlation between asymmetries and physical defects. Symmetry is also a relatively easy geometric feature to encode in optical receptors, so the two go together. There is also a universal preference for clear skin, and this seems similarly related to general health. Some skin blemishes, such as acne, are not signs of ill health, but so many are that this looks like an easy proxy for good (as in survivable) genes. As before, it's not that we are attracted to clear, symmetrical faces because we are looking for good genes. Rather, we have come to like clear, symmetrical faces because the genes that lead us to prefer them in mates had a better chance of getting into the next generation, since they were hitching a ride with genes possessing better fitness.

The few universals that are gender specific appear to be proxies for youth and fecundity in women, and strength and power in men. Men prefer women with higher cheeks, larger eyes, smaller noses, smaller chins, fuller lips, larger breasts, and lower waist to hip ratios (the hourglass figure). The facial features are all characteristic of youth, which is a proxy for fertility. Women prefer men with square jaws, commanding expressions, deeper voices, and athletic bodies. Grey hair is perceived as less attractive in women, but neutral to attractive (in moderation) in men. This is a straightforward proxy for age and thus lessened fertility in women. But it is neutral for fertility in men, and increased age is a partial proxy for power and influence. Again, we are not generally aware of the features our attraction standards select for, such as fecundity, or even the signals of those features, such as age. We have an innate metric of attraction that orients us more toward certain individuals than others. We are attracted to those individuals because we perceive them as more beautiful or more handsome. We have those innate perceptions because the genes that encode them rode along with the genes for which they are proxies.

Curiously, one feature of our human concept of love appears to have a better basis in our ancestors than in us. Our literature and culture are filled with the notion of "love at first sight." It is a romantic notion that we prefer in our stories and imaginations, but are skeptical of in actual practice. A companion concept is "the one," the ideal mate that is somewhere out there to be discovered – someone so perfectly fitted to us that we were fated to be paired. If you manage to encounter this someone, you recognize each other via love at first sight – it's you! All of our many failures to sustain a lasting relationship are chalked up not to our lack of skill in relationship building, but to having found the wrong one. It takes language and culture, and career goals, and expectations, and introspection, and endless dithering about whether we made the right choice to unwind a relationship that first begins with the addictive rush of initial attraction. Our less cultured cousins in the animal kingdom don't have this problem. If they are a pair bonding species at all, the natural interplay of dopamine and oxytocin converts the initial attraction into a long-term bond. Love at first sight works out every time because the first choice always *becomes* "the one." The fates don't pre-pick the ideal pairings and then torture the erstwhile lovers with the improbability of ever meeting. We, on the other hand, are unable to sit back and let nature take its course because we have a unique fourth emotional system that plays havoc with the other three. We have a compelling need to belong to, and be accepted by our surrounding social groups – peer, family, cultural, professional, state and religious. And our love lives and mating choices have come to be regulated by those groups.

Rules of Engagement

It is ironic that we evolved these emotions for romantic love as strategies for reproduction and gene preservation, in a context in which we were consciously aware of neither of these goals. These emotions were, for our ancestors, ends in themselves. Now that we are able to consciously reflect on our romantic emotions, to talk about them, to write about them, to devise strategies to promote and retain them, we actually regard them as ends in themselves – as if this were the point all along. We don't essentially connect love to reproduction or gene preservation. And having thus reified love as its own goal, we then went on to build complex cultural and social systems to regulate love in strategies for – you guessed it – reproduction and gene preservation. Our cultural ancestors did not know much about genes, but they did know about reproduction. And the reproductive consequences of love were very much on their minds because it is only through properly indoctrinated progeny that social institutions can themselves survive. So every system of social organization that depends on cultural artifacts being faithfully passed down the generations asserted an interest in who mates with whom.

We tend to think of marriage as a relationship between two individuals, but it is primarily a relationship between two families and two estates. Social standing, property, religious beliefs, legacy, and inheritance are all at stake. That's why churches and states grant the licenses and officiate at the ceremonies. There are rules of engagement. A look back at recorded history shows how pervasive these rules once were. Tribes of all sorts have prohibitions on marrying outside the tribe. This is both a racial and cultural phenomenon, a way to preserve genes and memes. In the case of kingdoms, royalty had to be carefully interbred to preserve the "divine right" gene. Pursuit of this perceived gene often had negative consequences for the actual genes of monarchs because of the severe lack of genetic diversity. At the other extreme, it was often thought to be good practice for the conquering kingdom to rape and pillage the conquered, as spoils of war. An advantage of the rape part is that it spread the conquering genes into the conquered population, tribally annexing them so to speak. There is genetic evidence that about 1 in 10 men who currently live within the borders of the original Mongol Empire are direct descendants of Genghis Khan. That translates to about 1 in 200 men worldwide.

In most cultures, there were rules of engagement, explicit and implicit, that regulated marriage to preserve internal social hierarchies – peerages, classes, castes. Families had their own unwritten rules to use the marriage of their offspring to either maintain or improve their standing in the hierarchy, and certainly to avoid losing relative position due to an unfortunate union. These became codified in property laws and customs that regulated the exchange of sons and daughters, often requiring fathers of daughters to provide dowries to make the exchange equitable. Daughters could be exchanged for goods. The fine line between this and the lowly crime of prostitution was that the former was in the higher service of preserving family legacy. A very respectable strategy for ambitious families was to offer daughters as concubines to monarchs, gaining favor, or court access, or land, in the least case, and a potential promotion to queen, in the best case. In many cultures, marriage was (and in some, still is) explicitly negotiated and arranged by the families, independent of the wishes of the betrothed.

Religions inserted themselves in the process to preserve the all-important context of parent-child indoctrination. Early indoctrination is essential for maintaining allegiance later in life, and this is not likely to happen reliably with a mixed-religion couple. Churches managed to define marriage as at least a partial divine license that must be granted by a local Earthy authority on behalf of the Gods. This required even monarchs to negotiate with the Church to sanction their often capricious or political unions (and disunions). There is a great irony in the fact that Western religions often regarded sex as inherently sinful. It was tolerated for reproductive purposes – this being its "natural" purpose. The early theologians that established these doctrines could be viewed as the first theorists to grasp the essential facts of evolution. Sex is not there because it is pleasurable or desirable in itself, but because it reproduces people. They were the first to admonish us to keep our focus on the actual goal of the mating game.

In most of these institutions, romantic love was considered a dangerous force, something that easily gets out of control and leads to unions that muck up the proper social order. It is something to be controlled, to be channeled into appropriate outcomes. So there were all kinds of lesser rules of engagement concerning how men and women of reproductive age could and could not associate with each other. In general, unchaperoned association was a recipe for trouble, but if you didn't explicitly arrange the marriages in advance, you needed some safe way for courting to proceed, for families to be vetted, and for proposals to be considered. So each culture had its own customs for who called on whom, who had to be present, and euphemistic signs for showing and reciprocating intent without having to be so vulgar as to actually say so. This was the official, cultured view. Of course, running in parallel, the underground culture of brothels, gentleman's clubs, taverns, festivals, carnivals, midnight trysts, and drunken revelries continued to celebrate love for its own sake. We never lost track of the end-in-itself notion, it just had to operate at the margins of society.

And it's not as if the romantic concept had no high-minded culture of its own. Quite the contrary, most of the romantic literature of our past celebrates the fate, usually tragic, of lovers who have found each other at cross-purposes with their social context. It is clear that we, the audience, are supposed to revere the lovers and their higher calling, not the Montagues and Capulets, or the evil stepmother, or the Trojans and the Greeks, or the Earnshaws and the Lintons. If we are so attuned to the primacy of love over culture, at both the high and the low end, why do we so often kowtow to culture when it conflicts with love? Why has there not been more rebellion?

The reason, as we have noted before, is that we are an inherently tribal species. We have our own special class of social emotions that drive us to seek the acceptance of the tribe, and fear ostracism by the tribe. It is not hard to imagine why such emotions had adaptive value in our evolutionary past. Individuals that are driven to bond with each other in tribes can act as tribes, sharing resources, sharing child rearing, sharing defense, sharing offense. Unaligned individuals are at a disadvantage when competing with tribes. Tribes that bond loosely are at a disadvantage against tribes that bond strongly. We can measure the primacy of the emotions that underlie this drive with a simple experiment involving a game of catch with balls. The experimental session is staged as an activity between three people who will alternately throw and catch a ball among themselves. Unbeknownst to the subject who is being measured, the other two participants are working for the experimenters. After a few minutes of shared tossing and catching, the two confederates begin throwing only to each other. Only a few minutes of this exclusion are enough to generate strong negative emotional responses of anger and sadness from the excluded player. This same effect can be measured when the ball throwing game is virtual, run on a computer where the subject is told they are playing remote players on distributed computers. It can be measured even when the subject is told that the other players are members of some reviled group such as the KKK.

And finally, it can even be measured when the subject is made aware that he/she is playing a computer, not real people.

It is not a rational response; it is an emotional response. The brief period of cooperation stirs the emotions of group inclusion. The emotional brain gets into "the zone," similar to the willing suspension of disbelief that you enjoy when reading a novel or watching a movie. The sudden ostracism occurs in the context of this virtual reality, raising the negative reflex. When subjects undergo fMRI brain scans while playing the computer version of this game, the neural pathways that light up at exclusion are the same as those involved in physical pain. It really hurts to be ostracized.

What makes this particular emotional system so potent is that it incorporates a degree of indirection. The romantic reaction to an attractive partner is specific to physical features of that partner – the same kinds of features that caused the attraction to evolve in the first place. The tribal inclusion emotions evolved to favor *any* tribe that would take you. How that tribe identifies itself, what it believes, what it asks you to believe, what it asks you to do, what it asks you to refrain from doing – all of these things are variable. You do not necessarily like what your tribe does, and you may love what your tribe prohibits. You are responding to the generalized need to be included, so this can trump other emotional inclinations. This retargetable dynamic is what enables extraordinary acts of courage and self-sacrifice in time of tribal warfare, from otherwise cautious or timid people, as well as unconscionable acts of cruelty from otherwise respectful and empathetic people.

The generalized emotional systems for attraction/addiction and pair bonding became romantic emotions by being linked to the mating system via physical evolution. The tribal-inclusion emotions became linked to our love life (and everything else) via cultural evolution. When our social tribes added rules of mating engagement to the inventory of things members in good standing obey, we gained this fourth dimension to our romantic motivations, one that works more often than not at cross-purposes with the other three. We are not as tightly regulated as we once were by these rules, and we like to think of ourselves as somewhat immune to this kind of influence, but it is still there in the background, particularly in the context of families. Women, for instance, who feel no specific inclination toward having children, either positively or negatively, still report the impending sense of their "biological clock" counting down as they enter their 30s. Families expect children. Friends are having children. It is what normal humans do at this stage in their lives. There is an urge to belong, to take your place in the social structure. This begins to limit mate choices to those that will be compatible with families. Differences in ethnicity, religion, race, age, or socio-economic status that do not affect individual attraction suddenly become relevant. Promising relationships are ended by both genders when the thought of taking this person home to meet the family arises.

Private Spaces

What about our most unique romantic trait: our sense of privacy and intimacy in love? There are no (other) animal models for this. Every human culture has a version of this. Some cultures are more modest than others about the amount of clothing that is appropriate in public places, but we all have the notion of private parts, private spaces, and private feelings. Intimacy is about sharing these private things with a special confidant. It is about shared secrets, something that sets the two of you apart from everyone else because you have implicitly agreed not to disclose what you share.

It might be natural to assume that this kind of intimacy is a cultural artifact, that we learned this kind of behavior out of a need to keep romantic bonding out of sight of the public rules of engagement. Because of the inherent conflict between love and social obligation, there was always a need for this kind of privacy, and certainly secrets to be kept. This accounts for the "us against the world" point of view that makes you co-conspirators in conflicts with your tribes, but it doesn't account for the earlier, more difficult conflict in negotiating this intimate relationship between yourselves. This conflict is more basic. It is part of our innate social emotions. It is not unique to humans – other highly social primates exhibit this – but it is greatly magnified in humans by our capacity for language and culture.

Living in groups always presents conflicts between the personal desires of individuals relative to each other. There are resources to be divided, alliances to be made, conflicts to be resolved. Species without strong socialization simply fight it out on a case-by-case basis. Each interaction is a new encounter. But if you are a social species, you have a more nuanced understanding of the state of mind of your adversary. So you play a kind of poker where you try to infer your adversary's true motivation while keeping your own hidden, or perhaps intentionally deceive your adversary as to your true motivation by bluffing. Your opponent, of course, is pursuing the same strategy, so you have to be on guard for deception as well. Trust does not come easily in this kind of engagement. In fact, trust is just another bargaining chip. You want your adversary to believe they can trust you, and to believe you trust them, even though neither may be the case. You also know you are likely to encounter this individual again. Others will be watching. Your actions will contribute to your reputation, and will affect how others deal with you later.

All of this contributes to the need for two personalities: your private persona, carrying your true feelings and desires, and your public persona, an actor that plays the social games and carries your public reputation. Other primates do this to some degree, but we are the masters of manipulation and deception because we have so many more dimensions for informing and misinforming via language. We rarely say what we mean. We say what we want others to think we mean. This is both innate and learned behavior. The innate tribal need to be accepted, and the fear of rejection, drive us to build up a public persona that will be accepted. Children are initially very candid and trusting, expressing the inner self directly, but gradually learn how to behave and what to say (and not to say) to build a successful public persona around the private one. Trust then becomes an earned commodity, not a free one. This is why we innately admire people who speak their minds, even when we disagree with them, and innately distrust politicians and lawyers, even when we agree with them. We don't like to be manipulated, even though we find it necessary to do it ourselves.

This sets up an inherent contradiction in the quest to find an intimate soul mate. Two public personas must engage in a delicate social negotiation whose ultimate goal is to expose two private personas. Some of the emotional goals of love – sex and early attraction – can be obtained by our standard manipulation strategies. So there is a temptation to stay public as long as feasible while trying to learn as much as possible about the private persona of your date. Perhaps all of the time-honored dating strategies, such as playing hard to get, or trying to maximize your mate-market value in the exchange, involve manipulation. There are winners and losers. This competes directly with the non-zero-sum outcome of intimacy, where both parties receive optimum benefit. To achieve intimacy, each party must allow the other into their private space, emotional and/or physical. This requires simultaneous trust, but trust is almost never simultaneous. Someone has to go first, either to unilaterally enter the other's space or to unilaterally permit entry into their own. There is great social and romantic risk in not being reciprocated. We tend to overlook how our endearing, refined concept of love is intertwined with vulgarity. The same act can be seen as tender or vulgar, depending on the degree of intimacy that is already in place (or not). Timing, coordination, mood and context are everything. We can only achieve intimacy when, through some miracle of timing, both parties agree to give up their public facades and connect directly with their private personas. Our animal cousins don't have this extreme public/private dichotomy, so their pair bonding doesn't require them to achieve a shared privacy. Perhaps our bonds are deeper because of this. They are certainly harder to achieve.

Oxytocin helps. It lowers social inhibition, raises trust, and skews your perception of mate-market value. It almost seems like nature's ideal solution to our culturally induced intimacy impasse. But it was already there. We managed instead to dull its effect by outsmarting ourselves with clever social strategies. We write about this all the time. The happily-ever-after love stories are relegated to the pulp romance genre. The more refined, literary fiction of love deals primarily in the ironies and failures of earnest lovers missing the inflection point of intimacy despite trying heroically. There are so many ways to fail, and so few to succeed. We pursue love like we pursue happiness in general. Because, unlike other animals, we can plan ahead and imagine future emotional outcomes, we are painfully aware of what we don't have in the present. We are perhaps the only species that can feel sad just because we are aware that we are not more happy. We know what we are missing. So we are driven by strategies to get us to this better place. And we are notoriously bad at this, both in predicting what will get us to where we want to be, and in what we will feel like when we get there. So we achieve fewer long-term pair bonds than our cousins who don't know what they are doing.

It's hard to imagine that this impossibly complex kind of love is some kind of evolutionary advantage. It is the result of evolution, both physical and cultural, but that doesn't mean it had to be a good idea on its own. It is more likely a neutral side effect of the good ideas of gene preservation and social cohesion. Evolution only guarantees that you will survive, not that you will be happy. Happiness, and especially one of its most cherished forms – love, we are left to figure out on our own.

14 | Empathy

I do not ask the wounded person how he feels, I myself become the wounded person.

– Walt Whitman, *Song of Myself*

Human empathy seems like some sort of miracle. It is our noblest trait, some would say, disposing us to favor others over ourselves. We celebrate altruism as the exemplar of ethical behavior, but altruism is *just* a behavior. You can behave altruistically because you believe you are supposed to, or because this is what society expects of you, or because this is what your religion requires of you, or because you will feel embarrassed or shamed or shunned by the group otherwise. You can do and feel all of this without feeling any empathy for the person you are behaving selflessly toward. You may not even like this person. Altruism in the absence of empathy can be explained as self-interest. Your tribe will exact a cost for your selfish behavior, so you are behaving altruistically to avoid a cost to yourself. Empathy, on the other hand, drives you to incur a cost to yourself for the benefit of someone else. There is no anticipation of reward. You experience some sort of out-of-body bonding with another person that aligns your interests and makes their welfare important to you. You want to do this for its own sake. How does that possibly happen?

Even biologists struggle with this phenomenon. How could such an innate motivation have evolved by natural selection? Preserving your genes requires vigilant self-interest. You can't, in general, afford to incur costs to promote the genes of someone else. That kind of behavior will not get passed down the generations. We can explain the selfless behaviors toward lovers and children and grandchildren because they all carry your genes, so by helping them you are helping your genes, if not yourself. Biologists typically explain the extreme, self-sacrificing social behaviors of insect societies, such as ants, via this same relatedness relation. It's called *kin selection*. Behaviors that promote your genes, whether the genes are in you or in your kin, will be selected for. There may be cases where such behaviors kill you off to make it more likely your kin will survive. This works. But altruism, and the native empathetic emotion that promotes it, is directed at arbitrary individuals who most likely have no genetic relation to you, even complete strangers. You are promoting the welfare of someone else's genes at the expense of your own. That's not supposed to work. That kind of behavior gets snuffed out.

If you believe in miracles, or perhaps just hope for miracles, or more specifically, hope that some of the miraculous-seeming elements of our human nature remain forever immune to biological reduction, maybe this is where you can take your stand. Go ahead! Reduce this, why don't you! Well, it turns out it's not a miracle after all. It was just a tough nut to crack. Now we think we've pretty much got it figured out. And it's absolutely fascinating.

It's Not Rational

On the face of it, evolving an empathetic emotion that promotes kindness to strangers seems impossible. Unilateral kindness promotes unrelated genes over yours, both because you give up a little of your own resources, and you bestow a benefit on someone else. If the stranger has selfish genes, your kindness genes just make selfishness more likely. So in a population full of selfish genes, a new mutation that gives an individual a kindness gene will go nowhere. Kindness genes need some returned kindness in order to compete. For everything you give up, you need someone to repay the favor. This poses serious problems for ever getting started in a selfish world. So it seems you need some kind of critical mass of empathetic individuals in order for them to establish a beachhead.

This raises the idea of group selection. If a group of mutual-aiders can manage to get started at all, it seems that they should do better jointly than unaligned, selfish individuals. So it seems you need two things to evolve empathy: some kind of scaffolding that keeps a fledgling group of empathetic individuals together until they reach critical mass, and some group benefit which emerges from their cooperation that enables them to outcompete selfish individuals. Sounds plausible, doesn't it? But there is a problem with this story. It still doesn't work, even if you get a protected head start, and even if the benefits of mutual aid are greater than the benefits of going it alone.

Consider the following scenario from evolutionary game theory, known as the *Prisoners' Dilemma*. Two suspects accused of a joint crime are each offered the same plea deal by the district attorney, but they are isolated in separate rooms so that they can't communicate with each other. If the suspect confesses and testifies against the partner, he will go free and the partner will get the maximum sentence. If neither confesses, the DA's case will be weak, so they will both likely get short prison sentences. If they both confess, the DA's case is made, so they will both get substantial sentences, but less than the max because of their confessions. The DA is hoping that both will realize that the rational choice is to confess according to the following reasoning. If the other suspect confesses, you will be better off confessing as well so that you don't get the maximum sentence. If the other suspect remains silent, you are also better off confessing because you will go free. This drives both suspects to the suboptimal outcome (for them, not the DA) of a long sentence. They miss the better joint outcome because neither can trust the other to remain silent. It would, after all, be irrational. We don't know what Captain Kirk would do in this situation, but Mr. Spock would confess.

The intuition behind group selection is that the higher benefit (lesser punishment) of the two prisoners cooperating with each other should, over time, outcompete the lower benefit (greater punishment) of each looking first to his own self-interest. After all, a population that learns to trust each other will enjoy a greater average fitness than a population that cannot. So it seems if you could just get enough of a mutual trust regime going to get over the hump, by some means or other, group selection would take hold and cooperation would be selected for. The problem is that the greatest benefit of all goes to a self-interested individual that exploits a cooperator. So there is always a higher incentive to cheat.

Economists struggle with this dilemma as well because it implies that rational self-interest will always result in a population jointly choosing the less optimal joint economic outcome. How does cooperation ever get started? Arguments come and go about how individual confrontations between pairs of people of the Prisoners' Dilemma variety will evolve into population-wide social structures in the long run, but it is just too hard for humans to predict such outcomes. That's why we have mathematics.

At the intersection of biology and mathematics there is a field called *evolutionary dynamics*. Evolution by natural selection is a generic process that can occur among many kinds of phenomena, not just biological ones. Languages evolve, cultures evolve, ideas evolve, social networks evolve. And evolution occurs at more than just the organism level in biology. Genes evolve, viruses evolve, cancers evolve, immune systems evolve. Any process that has entities that both mutate and replicate in an environment that causes some kinds of those entities to replicate faster than others will exhibit evolutionary dynamics. Those dynamics can be studied mathematically, either in the abstract to prove that certain laws and relationships will always hold for problems of a certain type, or as models of natural processes to predict likely outcomes over time, or to explain under what circumstances particular outcomes are likely to occur. The problem of cooperation, exemplified by the Prisoners' Dilemma, is studied by investigating the dynamics of evolutionary games. These games model the competition between individuals to leave more of their progeny in the next generation. The dynamics of such games can illustrate the relative reproductive advantage of two different kinds of individual that encounter each other repeatedly in a population, and thus predict what kinds of population structure will emerge over time.

The difference between two kinds of individual under study is expressed as a difference in the strategies that they use to play the competition game. As with the mating strategies we looked at earlier, these strategies do not need to suggest that the entities playing them are conscious of the strategy. They are simply shorthand for predictable behavioral dispositions that the players bring to the encounter. These could represent actual conscious strategies of humans in social interactions, or the unreasoned emotional responses of dogs, or the molecular cascades that link input to output in bacteria. In fact, at the level they are studied in evolutionary games, strategies are defined simply by the value of their outcomes against other strategies. You don't even have to say why this occurs. The outcomes for all encounters between two kinds of strategies, A and B, in a population, for instance, are given by four values (known as a payoff matrix): what an A scores when it plays another A, what an A scores when it plays a B, what a B scores when it plays another B, and what a B scores when it plays an A. The scores represent relative reproductive success. A strategy with a higher score will leave more progeny of its kind in the next generation. This will cause the ratio of As to Bs to change over time in favor of the better strategy.

Sometimes you can completely characterize what will happen over evolutionary time from the structure and relative scores of the individual game itself. In this case, a differential equation can be derived that describes all possible evolutionary outcomes. Sometimes the dynamics are just too complex to be analyzed by solving equations on a chalkboard, so you have to resort to simulating them on a computer. Even then, the combinatorial complexity of the model may be too great to compute exhaustively, so you settle for simulations of representative outcomes from random starting scenarios. The outcomes can vary from the eventual total domination of one particular strategy that drives all others to extinction, to stable equilibria where several strategies eventually settle into a ratio of mutual coexistence, to oscillating, out-of-equilibrium patterns, to chaotic effects that yield no predictable patterns. Sometimes the outcomes depend on the initial ratios of the competing strategies, and sometimes they don't.

The Prisoners' Dilemma game is of such interest, to biologists, mathematicians, economists, and business schools, because it turns out that the scores for each of the four outcomes don't really matter as long as the *relative* scores are in a particular order. It is a model for what happens in all games where individuals with cooperative strategies (or genes, or dispositions, or traits – such as empathy) encounter individuals with freeloader strategies (or genes, or dispositions, or traits – such as selfishness) with the following relative payoff matrix. The highest score goes to freeloaders when they play against cooperators (the freeloader participates in the group benefit without contributing any group cost). Second place goes to two cooperators playing each other (they share in the group benefit and split the cost of obtaining it). Third place goes to two freeloaders playing each other (they pay no cost, but get no group benefit either). Last place goes to a cooperator playing a freeloader (the cooperator shares the group benefit with the freeloader, but pays all the cost). (Some theorists require additionally that the score of $2^{nd} > (1^{st} + 3^{rd})/2$, but we won't worry about that).

For this class of games, the results are definitive and don't depend on the initial ratios of cooperators and freeloaders. Freeloaders always dominate and eventually drive cooperators to extinction. The freeloading strategy is what is known as a *Nash equilibrium* (after John Nash, Russell Crowe's character in *A Beautiful Mind*). If two players adopt the strategy, neither can improve its outcome by switching to the cooperator strategy. Natural selection never manages to find the more optimal average fitness of mutual cooperation, but instead moves ineluctably to the lowest average fitness of mutual freeloading. So simple, unconditional cooperation, the kind of behavior that empathy motivates, is not only irrational, it can't evolve, no matter how much head start you give it. The empathy mutation gets wiped out soon after it appears in a selfish population. And even a completely empathetic population (however we first got there) is always in danger of collapse as soon as the first selfish mutation occurs. Empathy just can't compete. Neither Spock nor Kirk can get there.

Here natural selection is a proxy for game-theoretic rationality, choosing the "rational" outcome of self-interest even when the entities involved can neither think or feel. It is interesting that humans tend to follow the "irrational" strategy of mutual cooperation in experimental simulations of these games, reflecting the fact that our complement of community emotions has found something closer to the optimal group outcome, and thus that their evolution cannot be explained by the spontaneous emergence of empathy.

A Little Punishment Goes a Long Way

But we know we got there somehow, so how did this happen? Well, it turns out that you can't evolve kindness until you first evolve punishment. You need more than just a mutation that disposes you to cooperate; you also need a mutation that disposes you to do something about cheaters. Your cooperation behavior needs some additional mechanism for detecting and marginalizing non-cooperators. There are lots of actual ways that organisms can do this, but this more sophisticated kind of cooperative behavior can be modeled abstractly in evolutionary games with what are called reactive strategies. What the strategy does in any given encounter is a function of what happened in its previous encounters. In the simplest case, the strategy has a memory of only the most recent encounter. But this is enough to allow you to classify another player as a freeloader and thus do something about it in the next round. In this simple model, the only thing you can do about freeloading is to withhold your own cooperation the next time. This can be viewed as a sort of punishment, either retributive, if it is emotional, or as an incentive to get the opponent to cooperate, if it is rational. It could also be viewed as a purely defensive response. You don't know whether you can affect your opponent's behavior or not, but you've learned not to rely on it. At the level of the model, it really doesn't matter what your motivation is, or whether you are the kind of thing that even has motivations. All of these possibilities are covered by enumerating the range of possible reactive actions. Since there were four possible outcomes to the most recent game you played (you cooperated, your opponent didn't; you both cooperated; neither of you did; your opponent cooperated, you didn't), and you can do one of two things in reaction to each one of them (cooperate or not this time around), there are 16 possible reactive strategies. The two simple strategies, always-cooperate and always-freeload, that evolution resolves in favor of freeloaders in the Prisoners' Dilemma games, are among these 16. In the first case, you react with cooperation for all four previous outcomes, and in the second case, you react with freeloading for all four. You, in effect, don't really react at all.

When the field is expanded to include all 16 strategies, perpetual freeloaders do not always win. The clear winner is a conditional cooperation strategy called tit-for-tat (TFT). It starts out by cooperating, and then does whatever the opponent did in the last round. So against cooperators, it will continue to cooperate. It will stop cooperating if the opponent stops, but it will resume again if the opponent does. So it protects itself against exploitation by freeloaders, but gives them an incentive to cooperate by resuming cooperation if they will. Viewed anthropomorphically, this strategy is vigilant against cheaters, but it is not a slave to its emotions. It retaliates in order to incent cooperation. It doesn't hold grudges. It will always reward cooperation, no matter how long in coming. Stripped of these anthropomorphisms, TFT simply describes any cooperative strategy that maintains its fitness by cooperating with all and only other cooperators, whether the entity employing the strategy has intentions or emotions or neither. But it does describe the relative evolutionary value of an organism driven by a balance of retributive and hopeful emotions. In a population of unconditional cooperators, it is behaviorally indistinguishable from them. In a population of unconditional freeloaders, it behaves just like them after the first encounter. This gives unconditional freeloaders a very slight edge, so TFT cannot get started in an already established population consisting entirely of freeloaders. But a small proportion of TFTers in a mixed population will eventually drive unconditional freeloaders to extinction. In a mixed population with unconditional cooperators, TFT's conditional behavior is never exhibited, so everybody cooperates all the time, and there is no selection. Everyone has the same fitness.

These 16 possible strategies exhaust the deterministic reactive strategies – the ones that always react the same way to the same previous outcome. This makes it easier to compute the evolutionary dynamics, but such strategies are not a very realistic model of the social behavior of biological entities. In the particular case of emotion-driven behavior, we know that organisms that have them are typically driven by a range of competing emotions. No one behaves the same way every time. To the extent that organisms have a specific personality type, it is a tendency to behave in certain ways in certain situations. The characteristic behavior doesn't always happen. This can be due both to the complexity of interactions among the emotions, and to the possibility of mistakes, or weakness of will, in carrying out the action. All of these contingencies can be modeled by making reactive strategies probabilistic instead of deterministic. Whether the strategy will cooperate on the next move is given as a probability of cooperation for each of the four possible outcomes from the previous move instead of a yes/no decision. The 16 deterministic strategies are included among the probabilistic ones by making all of their probabilities either 0 or 1.

This creates a much larger space of possible reactive strategies, in fact, infinitely many. It is still often possible, though, to compute a complete characterization of the competition between any two of them. One result of exploring this larger strategy space is that another conditional cooperation strategy emerges that outcompetes TFT. This one is called generous tit-for-tat (GTFT). Like TFT, it always responds to cooperation with cooperation, but unlike TFT, it responds to freeloading with cooperation approximately one out of every three times. Anthropomorphically speaking, it has a degree of forgiveness. It is willing to give a freeloader another chance once in a while. This exposes it to more exploitation than TFT, but this is more than made up for by its ability to quickly recover from mistakes. When two TFTers play each other, and one mistakenly fails to cooperate once, the other switches to freeloading. This starts an alternating series of retaliations and resumptions of cooperation. If a second freeloading error is made, the two players become locked into mutual freeloading unless a further cooperation mistake breaks the cycle. Because of the forgiveness factor built into GTFT, a mistake will be corrected in an average of three moves, so mutual cooperation resumes. The ascendancy of GTFT over TFT shows the downside of an inflexible retaliation strategy. Such strategies are optimal only in a perfect world, where everyone is rational and no one makes mistakes. In a less perfect world – the one we live in, for instance – a modicum of tolerance produces a better total outcome. But because of this tolerance, GTFT doesn't fare as well against a population of unconditional freeloaders, so it is less likely to be able to establish the initial beachhead of cooperation. TFT is better at this.

Ten Million Generations of Evolution

Studying the relative strengths of reactive strategies in pairs doesn't tell us much about how stable cooperation might evolve out of self-interest from a standing start. To study this, we must consider random mixes of strategies competing against each other over time with periodic random mutations creating small amounts of new strategies. Given that the strategy space is theoretically infinite (and still prohibitively large if you impose an arbitrary precision limit on the probabilities) this must be done by representative computer simulation of real evolution. This has been done, with some provocative results. Simulated social evolution starts with a population all pursuing the maximally random strategy. Everyone will cooperate half of the time on all four outcomes from the previous round. Although conceptually odd, this does approximate the point in social evolution when normally self-interested individuals are first experimenting with cooperative behavior. No one has a coherent strategy for dealing with, or even recognizing, cooperators at this point because the behavior is fairly novel. Both potential cooperators and potential freeloaders need to develop appropriate behavioral responses. Every 100 generations, on average, a small amount of a randomly chosen strategy is added to the mix to simulate the effect of random mutation. When the concentration of an existing strategy falls below a small threshold value, it is eliminated from the mix. This is extinction. Each run of the simulation was allowed to go for 10,000,000 generations (thus looking at the emergence of about 100,000 new strategies by random mutation). The results show some clear trends.

Strategies whose probabilities approximate unconditional freeloading typically wipe out the more cooperative strategies and dominate the population for the first 100,000 generations or so. Then balanced cooperative-retributive strategies very much like TFT invade the freeloaders and catalyze a cooperative environment in which the more forgiving GTFT takes over. This creates a relatively exploitation-free environment in which unconditional cooperators can survive. Over time, the GTFT population will drift toward unconditional cooperation. When their retributive tendencies have sufficiently atrophied, unconditional freeloaders again invade, dominate, and the cycle starts again. This cycle of emerging and receding regimes of cooperation is eventually interrupted by a different variety of cooperative strategy, known as win-stay-lose-shift (WSLS), that is able to resist the random drift toward unconditional cooperation and thus keeps its brand of cooperation going indefinitely. At first glance, WSLS is a little unintuitive. It differs from GTFT in that it (nearly) always cooperates after a round of mutual freeloading, but it (almost) never cooperates if it was previously exploited by the opponent. If GTFT can be said to forgive every third exploitation, WSLS can be said to be unforgiving, but very hopeful. It is still strong enough, however, to resist invasion by unconditional freeloaders. It also differs from GTFT in that it will continue to freeload against an opponent if it has successfully exploited that player in the previous round. Since, like all cooperators, it follows mutual cooperation with more cooperation, this can't happen in a population of unconditional cooperators unless WSLS makes a mistake. If it freeloads by accident against a cooperator and discovers that there will be no retaliation, it continues to exploit the cooperator. Anthropomorphically, WSLS can be viewed as a hopeful cooperator with a touch of cynicism. If it happens to discover someone who never retaliates, it doesn't give the sucker an even break. This is what enables it to resist random drift toward unconditional cooperation like GTFT. So once it becomes king of the hill, it stays king of the hill.

Since the simulation uses random mutations, it is necessary to run the simulation many times to determine if a given outcome is stable. Although there was some variety in results between the runs, there was a clear trend toward cooperation in the long run. It had emerged in only 27.5% of the runs after 10,000 generations, but in 90% of them by 10,000,000 generations. In 82.5% of these, the final dominant strategy was some form of WSLS. What does this say about the actual evolution of our social emotions? What the results tell us is that given enough time, an evolutionary contest between a variety of cooperating and freeloading behavioral tendencies will be resolved in favor of a particular kind of cooperation, one that is retributive, forgiving, hopeful, and cynical in just the right balance. We can certainly recognize all of these traits in our current emotional mix, though we are far from a homogeneous population. We seem to have lots of cynics, naive cooperators, freeloaders, punishers, optimists, and hopeless romantics in various concentrations. Also, emotions such as empathy come in degrees. Some individuals are more empathetic than others. And the cooperative behavior of any one individual is a result of resolving conflicts among several competing emotions, empathy being just one of them.

We also don't know how long this is supposed to take. There is no concrete notion about how much time elapses between model generations. The intervals were not based on any particular biological lifetimes. So we could still be in one of the early cycles of cooperation before WSLS has emerged. Or we may have permanently subverted the normal evolutionary process at some point by interfering with natural selection in the way humans with cultures and laws and governments are capable of. We enforce cooperation, and protect the naïve from exploitation with institutions now, so one-on-one encounters are no longer the main drivers of fitness. So maybe we are now off the cycle, but we don't know where we exited.

But we have been subject to evolution over many millions of years as social primates, so the model is an appropriate morality play for the origin of our social emotions. It is unclear what the final triumph of WSLS means for us, but the earlier part of the progression tells us that emotions driving unconditional cooperation, like kindness, forgiveness, and unilateral empathy toward non-relatives, are unlikely to emerge directly from a population of self-interest. Conditional cooperation with retribution must emerge first. Perhaps this is why we are so tribal, and the most powerful tribal emotions are negative, ostracizing us for not cooperating. This is what enabled us to realize the greater average fitness of the common good, even though purely rational self-interest could not. The tribal system of retributive justice then set up an environment in which forgiveness could emerge to produce better outcomes. And this is the environment that eventually supports the luxury of individual kindness.

I Feel Your Pain

This evolutionary scenario suggests that empathy is a rather fragile emotion, that it is easily snuffed out each time it emerges until the social conditions are just right. If so, why does the empathetic tendency arise to begin with? If it is such a fragile, point emotion, unable to hold its own weight, what evolutionary forces could possibly keep this fledgling emotion alive while it is waiting to be adaptive? Evolution has no foresight, remember. Well, it turns out that empathy is probably not just a point emotion. It is likely just one of the many facets of a much broader neurological mechanism that has adaptive fitness across the whole range of social behaviors. Empathy is a natural side effect of being highly social at all. Philosophers have struggled for millennia to understand such a moral emotion that favors others over one's self. You can understand the sense of group obligation, because failure to aid others in public circumstances causes you social pain in the form of ostracism. We understand why we are motivated to avoid things that cause us pain or to favor things that cause us pleasure. This is how motivation works. So how is it that we are motivated to avoid others' pain or to delight in their pleasure if we can't feel either of these ourselves? Well, apparently we can.

In the late 1980s, a group of neuroscientists at the University of Parma (Italy) was studying an area of the premotor cortex in macaques. This is an area involved in the planning of motor actions. They implanted electrodes in individual neurons to record under what conditions the neuron would fire. A number of neurons had been mapped by this technique to fire for such behaviors as grasping an object and bringing a grasped object (typically food) to the mouth. As often happens in science, a groundbreaking discovery was made quite by accident. None of the four researches involved can definitively reconstruct the details of the eureka moment, or whether there were perhaps several such moments, but the general story goes like this. In between experiments, while a female macaque was resting with an electrode still implanted from a previous experiment that had recorded firing when the monkey grasped an object, the computer suddenly registered a firing event in the monkey's brain when the *researcher* grasped an object. The monkey had not moved. She was instead observing the grasping event of the researcher and firing the same neuron as if she had done the grasping.

Neurons like this that fire both for your actions and the similar actions of others came to be known as *mirror neurons*, because they mirror the behavior of others in our brains as if they were our own actions. It was eventually determined that about 20% of the neurons in this area of the premotor cortex are mirror neurons. This phenomenon turns out to be a major aspect of how primate brains work. The reason we study macaque brains is to gain insight into our own, so the question immediately arose: do humans have mirror neurons too? You can't (ethically) do the same experiments on humans, so research soon moved on to a different paradigm using fMRI scans on humans to confirm that we too have areas of the brain corresponding to the macaques' that light up both when we perform actions and when we see, hear, or even hear about the same actions being performed by others. Because we can't measure individual neurons in humans, the name for this capacity has shifted from *mirror neurons*, to the *mirror neuron system*, or just the *mirror system.* fMRI studies indicate that the mirroring phenomenon in humans is more pervasive than just the premotor cortex. We appear to have the general capacity to unconsciously map the actions, or even the stories of action, of others onto our own real and potential actions, allowing us to feel what it is like to be this other person.

This phenomenon is one we all recognize when we reflect on our willing suspension of disbelief when watching movies or reading novels. We vicariously live the lives of the characters. The new insight is that just by watching others perform actions, athletes for instance, we are implicitly training our own motor nervous system to imitate those actions. This is a generalized system for learning by imitation. Actual practice helps reinforce the lesson, but you get a head start just by watching someone competent do it first. And not surprisingly, our emotional responses are tied into the system as well. Perception of the facial expressions of others that signal their emotions are mirrored in the premotor areas of our own faces and connect to the emotions that produce such expression in us. When we see others express pain or disgust or joy, we vicariously experience the same pain or disgust or joy. When we see others' skin touched by a feather duster or pricked by a pin, we feel the vicarious sensation ourselves. We can also anticipate the emotion that another will experience just by seeing the actions that presage it. It is as if it were happening to us.

The neuroanatomy of this mirror system is still not understood in any detail, but the existence of such a system ties together so many disparate phenomena that used to puzzle us: why babies can imitate facial expressions within hours of being born; how you find yourself executing reasonable approximations of moves of your sports heroes, even though you have never physically practiced them; how we seem to be able to read each other's minds; why we feel each other's pain and joy. Such a neural system makes so much sense for evolving social behavior. It is not specific to emotional empathy. It covers all aspects of tailoring our individual behaviors to match those of others. We can act as groups because we experience as groups. What a great idea! So empathy doesn't need special protection or circumstances to emerge biologically. It comes with the broader program of vicariously living the lives of others. Such a system is useful from the very beginning of conditional cooperation in order to recognize, remember, and infer the selfish or cooperative intentions of others. What the special protections do, perhaps, is allow empathy to evolve in greater degrees, so that it wins progressively more of the emotional conflicts in individual decisions. We know that some of us are more empathetic than others, even in the present. Perhaps more of us are more so now than we once were.

15 | Free Will

We human beings do have some genuine freedom of choice and therefore some effective control over our own destinies. I am not a determinist. But I also believe that the decisive choice is seldom the latest choice in the series. More often than not, it will turn out to be some choice made relatively far back in the past.

– Arnold Toynbee

The last, and most recalcitrant, vestige of the soul in a biological world is free will. This would seem to be the one aspect of disembodied agency that can't be explained by biological reduction, because to reduce it is to make it go away. If our introspective sense of being free to choose whatever next outcome we wish can be accounted for by underlying biological causes, then it must have been an illusion to begin with. You can't explain it, it seems; you can only explain it away. Love can still be endearing, irresistible, and romantic even though it has biological causes. It is still love. Empathy can be even more noble and uplifting in light of where it came from and how its underlying biology binds us together as a species. It is not only still empathy, it is empathy plus. But how can free will still be free will, if you take away the free part? This is the mother of all gestalt shifts. Unconstrained freedom of choice in our actions is an essential part of every concept we have about moral responsibility. Yet the notion that every event has a prior cause is essential to our very concept of science. We appear to be constrained by logic to pick only one perspective, but the denial of either seems absurd.

This biological gestalt shift is different from the ones we have been considering. It has nothing essential to do with current biology. It has exercised philosophers, theologians and scientists for millennia. And it is primarily a concern, even now, only of these professional thinkers. We don't reflect much on this dichotomy in our daily lives. It is ancient and abstract. The only thing recent biology contributes is to make the issue a little more concrete. We are getting a little more articulate about personal identity and the biology of conscious choice.

The conflict between free choice and predestination is not something you find only at the intersection of science and religion. It has been an often-occurring conflict at the intersection of religion and religion. Minds that are disposed to believe in Gods at all, it appears, are disposed to believe in predestination. This is part of the background story. The Gods have a plan for the universe and for me. Things that happen happen for a reason – God's reason. Both good and bad things are chalked up to God's will. If this doesn't make sense to you, that's because Gods move in mysterious ways. The minds of Gods are inscrutable. Sometimes this fatalism gets canonized as official doctrine, such as in Calvinism, even in Christian religions whose heavenly reward scenario depends on human free will.

This is a serious management problem for theologians, balancing the omnipotence of Gods with their granting humans a certain role in the order of Earthly events. Fortunately, they have a more flexible cosmology at their disposal than modern scientists. In a universe where Gods create the appearance of law-like regularities interspersed with the spontaneous, free actions of disembodied agents, the natural laws only hold up to the borders of human choices. Something anomalous and free then occurs in a soul, which in turn starts the natural causes flowing again. Gods only preside over the causal connections between free decisions, giving humans some individual responsibility for at least part of their destiny. This means that Gods can't just wind up the machinery of natural laws and let it run unattended. It requires an incredible amount of divine attention, and the ability to restart after anomalous interventions, but that's the luxury of religion. Gods can do the most amazing things. The downside is that you just can't predict much of anything if human actions are involved, because the causal chains keep disappearing into the local Twilight Zones of free agency. (Gods can still predict – they can do anything – but clergy are impoverished).

Because science is in the prediction business, it has to have more discipline. It has to eschew miracles and anomalies in order to predict at all. Science can never prove that the *entire* universe obeys regular laws, but science must assume it.

Ordinary folks manage to skate by with both notions. We acknowledge that you can't be held responsible for what you are powerless to change. We accept some measure of responsibility for what we consider our own free choices, and we require a similar standard of responsibility, and thus a similar degree of freedom, from others. We do, however, acknowledge that some actions of free agents are coerced – someone puts a gun to your head. We also acknowledge that we often do things out of habit, without conscious choice. We acknowledge that human actions can be predictable (election polling). We are aware that our bodies do things over which we have no control (flinching). We acknowledge that someone's disadvantaged upbringing can affect their later life choices, perhaps mitigating some of their responsibility. We consider certain persons mentally incapable of behaving morally. We don't think animals have enough freedom of choice to be subject to moral sanctions. In each of these cases, we imagine some clear distinction between the free chooser and the constrained chooser, and apply our ethics accordingly. What recent biology is contributing to the deliberation is, as usual, a blurring of the lines. We are getting more gradations where we have been used to having boundaries.

Freedom as Unpredictability

Popular "paradoxes" often overstate incompatibility. A true paradox implies some sort of contradiction – that something, *A*, must be both true and false at the same time. Since logic is a branch of mathematics (actually mathematics is an application of logic and set theory), we can detect true paradoxes by representing them formally, then formally deriving a contradiction. Most alleged paradoxes don't meet this standard. They make an unwarranted assumption without which no contradiction results. The "paradox" of time travel is one of these. Folks have assumed that time travel is impossible because if true, you could go back in time and kill one of your ancestors, implying that you would never have existed in the present to be able to go back. Yes, that's impossible. But you could go back in time and *not* kill your ancestors. You could choose to do only things that are compatible with your present existence. Some time travel scenarios are incoherent, some aren't. No contradiction. When Einstein first proposed his theory of special relativity, an objection was proposed in the form of the "twin paradox." The theory implied that if one of two identical twins were to accelerate away from the Earth and return, at near the speed of light, the returning twin would be substantially younger than its Earth-bound sibling. Well, yes. That's what the theory implies. And it turns out to be *true*. There was never a contradiction, just something unexpectedly strange according to the common sense of the time (perhaps this still seems strange to you now).

A similar kind of thing is going on with the apparent incompatibility of free will and determinism. Philosophers have sometimes proposed the following "paradox." If determinism is true, at some point in the future we will be able to hook up a person to some sort of brain scanner that can predict what that person's next decision will be. If the prediction is announced ahead of time, the subject can simply choose something else when the moment comes. Contradiction? Well, the scenario assumes two things that might not be true. First, that if every event has a cause according to the laws of the universe, we will someday be in a position to predict any future event from prior ones. But the universe could be law-like, and yet we still might never be able to predict some events with any degree of precision. We already have this problem with quantum mechanics. So the scenario itself may not be feasible. Second, supposing we were able to automate such a prediction accurately, what's to say that the subject wouldn't choose what was predicted? The prediction would have taken into account the preannouncement and the feedback loop to the subject. There is just too much thought experiment and too little physics, including the thought that we might ever achieve such predictive precision in physics.

What we need to achieve in reducing the introspection of free choice to biological causes is an explanation of why the choice is perceived to be uncaused by the chooser. The chooser could be wrong about this, particularly since the causes occur at a more granular level than consciousness. We would also like, if possible, to preserve the notions of moral responsibility that are a part of the free will phenomenon. Knowledge of the actual causes of free choice might help true up some of the boundaries that we have assumed at the macro level. Dropping the broader, metaphysical inference about indeterminism may make no practical difference at all. The situation could turn out to be just like what happened when we discovered we really lived on a sphere and not a plane – namely, nothing. The loss of one theoretical implication from the larger collection of concepts made no practical difference at all.

Personally, I like the unpredictability scenario – that freedom of choice is essentially the unpredictability of choice. Every event has a cause, but we are often not in a position to know what the causes are. This is particularly true with regards to our own mental states where conscious introspection does not give us much insight into the behavior of neurons. And the behavior of a network of 85 billion neurons is sufficiently complex that we are unlikely to be able to predict it from initial conditions with any precision, even if we employ supercomputers. This matches our current state of knowledge, and much of what we've learned about predictability in general. Predictability comes in degrees. There are some things we can predict with stunning accuracy, such as the trajectory of an interplanetary spacecraft, some things we can predict only approximately over short intervals, such as weather, and some things that we just can't predict at all, such as the metastasis of cancer. All of these things have deterministic causes, but our ability to know the causes well enough to predict outcomes is limited.

It may even be physically impossible to predict certain things. If we knew the current state of every subatomic particle in the universe, for instance, we could in principle compute any future state. But to do this, we would need some physical medium in which to represent all of the initial particle states, and some computing mechanism that would transform them into a representation of the next total state. That would take more than all of the particles in the universe. So the only feasible computer that can do this is the universe itself, with every particle representing itself. We wait for them to reach the next state, then read out the answer. But now we no longer have a prediction. The next state has already occurred.

So we can only feasibly compute future states by leaving out lots of initial conditions, and abstractly modeling many small states with a few large states. There are kinds of problems, like spacecraft trajectories, where this loss of precision is not a problem. There are others, like weather prediction, where it is. And there are many naturally occurring systems that we classify as *chaotic* because their dynamics are so sensitive to small variations in initial conditions, that small errors are magnified exponentially as time moves forward. The future states of such systems are essentially unpredictable for any other than very short time intervals.

And finally, there is the problem of measurement irreparably interfering with the phenomenon to be measured. We already know about this problem at the quantum level. Knowledge involves representation. In order to measure any naturally occurring event, we need to intersperse an artificially occurring event that induces some signal, such as photons, from the thing we want to measure that reliably represents the property we want to measure. When you get to a sufficiently small scale, we no longer have probes small enough to avoid interfering with the property itself. So the measurement disrupts the original event, leaving us unable to predict what would have transpired had we not interfered.

We already accept the notion that lower animals have less freedom of choice than humans, and that sufficiently primitive organisms have no freedom at all. We correlate this with our ability to predict what they will do under controlled conditions. The only thing missing from this continuum, as usual, is humans. We somehow think of ourselves as completely free rather than relatively free. We are just the *most* free of all the organisms. There is a very straightforward scenario under which free will evolved over time that parallels the end result posited by theologians with their human islands of uncaused causes. Life was originally pretty predictable because it was much simpler. So the causes leading up to the surface of an organism could continue to be computed as they passed through the organism and came out the other side. Then some organisms (the animals) began to evolve nervous systems. This created little islands of more-difficult-to-predict transitions through the organisms. Still possible, but harder. As nervous systems evolved into ganglia, and brains, and super brains such as ours, the dynamics of the neural networks became essentially chaotic. This is the practical implementation of free will. We don't have islands of uncaused causes; we have islands of unpredictable effects. Each time the causal chain goes through a human brain, we lose track of what will come out the other side. We have to wait for the choice to be made.

Leaving aside the ethical implications for a moment (we'll get to those), let's recount the perceived properties of free will. We introspect a choice that causes our action, but we introspect no cause for that choice. It begins with us, so to speak. We often can't predict what humans will choose before they make their choice. If we could predict such a choice, publicly, a human could simply choose something else. With free will as unpredictable will, all of these features are still true. First of all, we often can predict what humans will choose. Advertisers do this all the time. If your tribe believes X, you likely will too. It is at the fine-grained level that we can't predict. Will you choose the right button or the left button? Would you like fries with that? What about the hypothetical paradox of prediction? Won't happen. Recall that the super-predicting-brain-scanner has to reach a conclusion *before* it announces the choice. This means it could not have counted the public announcement event as one of the causes at that point. At best, it would have to issue a conditional prediction. If the subject is told he will choose A, he will choose B. To compute an unconditional prediction, we would need more than a brain scanner. We would need one of those mini-universe computers that measures all of the particles prior to the whole scenario, so that we can account for all of the causes, including the feedback loops. But those are just too big, and they interfere with the particles. And finally, what about the introspection of an uncaused choice? Well, this is actually an introspection of a choice, and non-introspection of a cause. This would be your experience if you were unable to introspect the actual cause. And that appears to be what's happening.

Windows on the Soul

We have looked at consciousness before in relation to our perception of the external world. Our perception of the internal world, introspection, is the same sort of consciousness only it's directed at internal states instead of external states. We don't experience external states as interacting molecules, and we don't experience internal states as interacting neurons. We experience both as large-scale, summed up abstractions of these smaller events. Just as our conscious window on the external world has moved up the brainstem to the neocortex as our brain became more complex, so has our conscious window on the soul. "We," the ultimate executives that see the colors, feel the pains, think the thoughts, and make the big choices for the rest of the body, are dealing largely with the conclusions that lower minions in our nervous systems have already reached without our involvement. Our conscious experiences and actions come late in the chain. So we shouldn't expect to experience all of the neurological causes of our conscious choices. We perceive ourselves as choosing spontaneously, as not being aware of any prior inclination until the spirit moves us, so to speak. This is actually a very apt metaphor—the spirit moving us. It expresses the scenario, in soul terms, where something else, the spirit, initiates the choice, then pushes us into executing it. We are essentially observers being carried along by the flow. We sense the choice just after it has been made and is in the process of being converted into action. We watch our free choices being made through our conscious window on the soul. If it is a spirit doing the moving, then the scenario still has the problem of uncaused causes. But if it is the subconscious parts of our brains doing the moving, then the scenario matches what neuroscientists have recently been able to measure.

The first experiments aimed at timing the self-perception of conscious choice in relation to the neural activities that execute them were performed in the early 1980s. Subjects were asked to spontaneously decide when to make a quick hand movement while their brain activity was being monitored. The subjects watched the second hand of a clock and were instructed to note the time that they first felt the volition to move. This was compared to the buildup of a *readiness potential*, a measure of activity in the motor cortex and supplementary motor area of the brain leading up to voluntary muscle movement. The readiness potential precedes the actual muscle movement and is believed to represent the brain's prior intention to move. The experiments revealed that the *conscious* intention to move occurs as much as a half second *after* the readiness potential occurs, suggesting that our window on the soul has a certain degree of latency, sort of like tape-delay on live broadcasts which gives censors a few seconds to bleep things out.

These results created a bit of a stir in the neuroscience community, and the philosophy and psychology communities. Scientists prove free will is an illusion! A lot of experimental studies followed attempting to replicate, improve on, or criticize these early results. The general trend, however, has been confirmation. One critical study suggests that the readiness potential may actually be a neurological *attention* event, signaling the brain's readiness to make a go/no-go decision rather than the choice itself, buying some time for conscious volition to be at least coincident with the brain decision. But the same timing result can be produced using the *lateralized* readiness potential, a measure of greater activity in either the right or left hemisphere of the brain preceding a movement of the opposite hand. In this experiment, subjects are asked to make a spontaneous choice of which hand to move. The left hemisphere controls the right hand and right hemisphere controls the left. The readiness potential is lateralized to the appropriate hemisphere before the reporting of conscious choice. Here it seems that at least the choice of hand must have been committed before the conscious perception of the choice.

The precision of the results has been criticized for the usual unreliability of human reporting of awareness times, but similar results have been subsequently produced using experimental designs that don't rely on subjective reporting. The results have been criticized based on the holistic nature of EEG or fMRI measurements, but they have also been reproduced using electrodes implanted into individual neurons. This latter procedure is a strict no-no for humans (as we noted in the mirror neuron studies), but sometimes experimenters get to piggyback on the surgical procedures of clinical patients, with their permission of course. Neurosurgeons often have to probe brain areas for responses prior to surgery, so they are in there anyway with embedded electrodes, so why not stick a few more in for the advancement of science?

It's not often that scientific experiments get to impinge directly on ancient philosophical issues, so there continues to be a lot of discussion about just what these scientists have discovered. It's not just scientists vs. philosophers; it's also scientists vs. scientists, and philosophers vs. philosophers. Framing the issue as whether free will has been shown to be an illusion is no more productive than fretting about whether baseballs, or colors, are an illusion. To reduce is not to eliminate; it is to explain. That we've recently discovered some timing and precision errors in our conscious perception of things relative to their actual causes is not news. Our inaccurate perception of free will is no more illusory than our inaccurate perception of visual scenes updated with delays by saccades, or the inaccurate non-perception of the girl with an umbrella (from chapter 7). These are also described as illusions, but useful illusions, not alarming ones. They are approximations of micro-phenomena!

People get all wrapped around the axle about this because we've been led to believe there is a paradox. These results should come as no surprise to anyone who has been paying attention. The interplay between conscious and unconscious decision has always been murky. We make many unconscious decisions in the brainstem over which we have no effective control (sneezing, flinching, recoiling from pain). We make many unconscious decisions in higher areas of the brain where we are wired for possible conscious override that we rarely exercise – whom to love, for instance, or whom to feel empathy toward. We make many unconscious decisions that are well within the scope of conscious deliberation, but aren't because we have made such decisions before and now rely on autopilot – habits. Neuroscientists haven't discovered any big truths here, just some very small ones – like the relationship between oxytocin and pair bonding. These are useful things to know. They explain how macro things work in terms of micro things. We are big, complex beings whose conscious attention has evolved to be used sparingly in situations of great danger or novelty. Our conscious deliberations are usually spread out over long time intervals during which we are only vaguely paying attention to the details. On the whole, we can be said to be consciously making those decisions, but only because we are consciously intervening now and then. More often than not, our consciousness is just watching the show.

E Pluribus Unum

This notion of being in charge of your decisions, mixed in with not being in charge, should be familiar to anyone who has found themselves taking actions they would rather not have. Your better nature is inclined in one direction and you find yourself going in another instead. You chose, but you didn't really choose to choose. It's complicated! And indeed it is. As we've learned more about how the brain works (and how sometimes it doesn't), we've discovered the incredible complexity of a system of parts, and subsystems, and interconnections, and overrides. Some of this is due to what might be considered "well designed" neural systems of checks and balances, but much of it is due to the repurposing and reuse of disparate parts by evolution. Our mental life is a walking contradiction.

At the most coarse-grained level of organization, we have three cooperating brain systems: the original vertebrate brain that has a wide assortment of autopilot programs for surviving and reproducing in a dangerous and competitive world; the limbic system that processes our memories and our emotions, tagging memories with emotive content so that we also remember what was pleasant and what was painful; and the neocortex that allows us to predict, plan ahead, and select strategies and courses of action that will take us to emotionally happy places and away from emotionally painful ones (we think). Each of these systems has interconnection into the others, allowing for assistance and override. In general, the basic vertebrate brain drives the body through its daily routines. The limbic system keeps track of what's old and what's new, and how these things make us feel. It can intervene in the autopilot programs with motivations that drive us away from anticipated painful places and toward anticipated pleasant ones. The neocortex sees the bigger picture. It tries to look ahead and reserves the right to intervene in either the basic programs or the emotional programs if it thinks it sees a better outcome than either of these systems is choosing. Each of these systems is trying to help us survive and prosper. The difference is in the amount of look-ahead each has (or doesn't).

Within each of these systems, there are many sub-regions that specialize in different parts of the problem and thus must coordinate their actions with interconnections. On top of all of this, because we are bilaterians, our brain comes in two symmetrical halves that, more often than not, redundantly do the same things in parallel. Each side has some specialization, which can vary depending on whether you are right-handed or left-handed, that typically requires complementary specialization from the other side. There is a huge bundle of nerve fibers connecting these two hemispheres, called the *corpus callosum*, which enables them to coordinate their actions.

On the face of it, it seems like a miracle that all of these interconnected subsystems manage to sum up to one coherent personality that we call ourselves. We mostly don't experience the conflicts and the lower level negotiations. But we are all familiar with the sense of conflict that occurs within some higher-level subsystems, such as our emotions. Our controlling motivation for any given action is a compromise between many competing motivations. The emotions themselves compete for dominance based on intensity, and they compete with the frontal cortex that is constantly trying to balance near term gratification against longer-term wellbeing. We only notice the conflicts between whole subsystems when something goes wrong – schizophrenia, bipolar disorders, multiple personalities. We tend to think of the "normal," unified mind as a monolith rather than as a jumble of many sub-minds, where a few bad connections here and there would expose its true schizophrenic character.

Perhaps the most fascinating division of our negotiated consciousness is revealed when the corpus callosum, the connection between the left and right brain hemispheres, is severed. This is sometimes done surgically as a last resort to control otherwise debilitating seizures in epileptics (though the procedure is less common now than it once was). Healthy patients, with no other neural deficits, continue to lead normal lives after the procedure, and are often indistinguishable from those with normally connected brains, except for some interesting anomalies that are revealed in laboratory experiments. Our nervous systems are cross-wired so that the left brain receives input from, and controls the right side of the body, and the right brain runs the left side. The two brains are each capable of running their own side of the body as if it were an independent person. They normally coordinate running one, integrated person by communicating across the corpus callosum. One side or the other is usually dominant. For most of us, the left brain is dominant, which is why we are right-handed. The ability to process spoken language is usually lateralized to the dominant hemisphere, so for righties, this is the left brain. For lefties, it is lateralized to the right brain only about 30% of the time, though.

Experiments with split brain patients typically present two visual stimuli, words or pictures, so that each stimulus is in only one field of vision – the right eye sees only one and the left eye only the other. The patients are then asked to either describe and/or draw pictures of what they've seen. In a normal person, the input from each side would be processed and shared across the two hemispheres, yielding a joint conclusion. But without the corpus callosum, a righty will not be able to describe the input from the left side because the left brain, which controls language, doesn't receive it, and the right brain, though perceiving it, is unable to speak. The left hand, however, can draw a picture of the left stimulus, or spell out a word for it with Scrabble tiles. The dominant side with the language control essentially becomes the external spokesperson for the whole person, and will do its best to construct a rational motivation for the behavior that incorporates unseen content from its blind side. For instance, when a picture of a chicken claw is projected to the left brain and a picture of a snow-covered field is projected to the right, and the patient is asked to select a pair of picture cards that best go with the picture just seen, the right hand selects a chicken (to match the left brain's claw), and the left hand selects a shovel (to match the right brain's snow). When asked to explain the choice, the left-brained spokesperson explains the obvious chicken-claw match, but for the shovel, it constructs an explanation about that being needed to clean out the chicken coop.

These two selves are normally integrated into a single self, but when they are not, it is tempting to ask, "which is you?" The point of view of the dominant, spokesperson half that explains what is going on without being privy to a lot of the relevant information is somewhat like the normal, integrated you when trying to explain your motivations and choices without visibility into subconscious aspects of your motivations and decisions. You are responsible for the behavior of the whole brain, but you are not in immediate control of many parts of that behavior.

Now we are in a position to consider the moral responsibility implications of free will. Our laws, and customs, and mores hold the whole person responsible for its actions. The implication is that you, the whole person, could have done otherwise. You are in charge of your destiny. These laws and mores have never really considered the lower-level, neurological causes of your actions. Unfree, constrained actions are deemed such only when the constraint itself is visible at top level, such as being tied up, or pushed, or being coerced with a weapon. But the mores certainly take into account the many unconscious aspects of our choices. When you act out of habit, you act unconsciously, but this does not excuse your responsibility. You were *caused* to behave this way, but these habitual causes were laid down by you in previous conscious choices. We also have the notion of negligence. You may not have intended the consequences of your action, because you didn't foresee them, but we have a standard for the amount of due diligence an ordinary person *should* have exercised. There are conscious choices you could have made beforehand that would have made you aware of the consequences. None of these requirements obliges you to access some aspect of your unconscious brain, rather you are obliged to make the best of the many prior episodes of conscious access you've had.

So the fact that our conscious choices are caused by unconscious, or subconscious, neural events brings nothing new to the table of moral responsibility. You are still responsible for the behavior of the whole person because there are plenty of conscious intervals, over the longer course of decision making, in which you can intervene. Think of yourself as being in the position of CEO of a large corporation. As chief executive, you are responsible for the behavior of the whole corporation, even though most of the constituent actions are executed by your employees. You are often not aware of their individual behaviors, or how the decisions of some of your middle managers are formed. The law does not hold you responsible for a murder committed by one of your employees, though it can hold the corporation liable if the murder was committed in the course of company business, because the company didn't exercise reasonable oversight. And the CEO can sometimes be held liable as well if the offending corporate behavior can be connected directly to irresponsible or negligent behavior on the part of the executive. A CEO knows this, and knows, that like the commander of a large aircraft carrier, there is a certain degree of latency between your issuing of steering commands and the ability of the vessel to execute them. These things turn very slowly, so you have to plan ahead and take this latency into account.

You, the conscious part of your brain, are the CEO of your body. You are responsible for how the actions of all of the many competing sub-yous get consolidated into a top-level action, because you are the top-most executive in control. You are unable to micromanage the behavior of your employees, so you have to set clear standards, encourage good habits, monitor your progress, see things coming, take reasonable precautions, know when to be paying close attention and intervening, and when to safely delegate to the middle managers. You have ample opportunities to steer the ship in the right direction, but these opportunities rarely come at the last minute (or within 200 milliseconds). This is the insight from Toynbee's quotation above.

Unlike a CEO, you can't hire and fire your subordinates. You are stuck with the subconscious brain regions and neurons that you have. So if their native instincts dispose them to inappropriate actions when you are not watching, you have to *train* them, and re-train them. This makes it much harder to be the morally responsible CEO of a defective corporation, where some of the employees are dysfunctional. Some of them may not be trainable. So we give you a little more slack if your brain is somehow defective, or missing something, or abnormally configured. There is always the ambiguity about whether you are trying hard enough vs. being physically unable to govern your body effectively. Neuroscience doesn't really change any of this. The moral responsibility aspect of free will already takes non-conscious causes into account. It already has the appropriate gray areas. So the traditional free will aspect of the meaning of life is one of the ones whose outer contours are left largely intact by the new biology. We are still as free as we ever were.

16 | Artificial Life

I saw—with shut eyes, but acute mental vision—I saw the pale student of unhallowed arts kneeling beside the thing he had put together. I saw the hideous phantasm of a man stretched out, and then, on the working of some powerful engine, show signs of life and stir with an uneasy, half-vital motion. Frightful must it be, for supremely frightful would be the effect of any human endeavor to mock the stupendous mechanism of the Creator of the world.

– Mary Shelley, *Frankenstein*

What I cannot build, I cannot understand.

– Richard Feynman

If God had meant us to fly, he would have given us wings. If God had meant us to create life, he would have given us – what? We weren't supposed to be able to do this. But as with flight, and any number of other "unnatural" capabilities, we figured out how to do it anyway. Artificial flight, though, didn't cross over any interesting theological lines. The holy books don't address it. It was merely a parochial skepticism of the times. The holy books don't really address artificial life either. It was never considered even a remote possibility, so there wasn't any need to prohibit it. Adam and Eve were thrown out of the Garden of Eden for being curious about things they shouldn't have. They ate from the tree of knowledge and learned the Godly perspective on good and evil. This was a bad thing, and it set the precedent for there being secrets of the universe that are reserved for the Gods. But you only have to prohibit humans from doing things they are actually capable of doing. What possible metaphor could have been used for Adam and Eve illicitly competing with God by creating some *de novo* life forms on their own? This would be like prohibiting humans from becoming omnipotent.

It wasn't until millennia later that we could even conceive of humans creating life, and when we did, it became the province of the mad scientist, replete with the moral tale of nefarious usurpation of the power of Gods. Mary Shelley's *Frankenstein* is the archetype: dark, foreboding, powerful, gruesome, involving the assembly of used body parts and some inscrutable "powerful engine" to provide the vital spirit of an artificial soul. No one had any idea how you would go about creating life, but if you did, it must happen something like this. This was still safely science fiction. But when we actually crossed over the line, it was very much like the initial abiogenesis event(s) that we described in chapter 3. A very small, incremental step that everyone involved expected to happen because the basic mechanics of life are now so well understood. Perhaps you missed it. It happened in 2010. It didn't make the nightly news because as biotech innovations go, it wasn't all that exciting. We've done much more provocative things with life than this in the last 20 years or so. This was more like dotting the last 'i' or crossing the last 't' – confirming something we always suspected.

Now that we've managed to usurp the power of Gods, what does this say about souls? Scientists still don't have a clue about how to make them, or how to get them into bodies (labs aren't really trying very hard). When souls were thought to be the vital essence of life, what made bodies living, this assured people, even scientists, that we could not create artificial life. We couldn't make the soul part. But we subsequently figured out how to make every *other* aspect of living bodies, and it turned out, in the end, that this was enough to create life. What does this mean? Well, there are three possibilities. 1) Perhaps lives don't have souls after all. If you manage to reproduce all of the more tangible aspects, you're done. 2) Gods (perhaps reluctantly) granted humans a creation franchise, and agreed to cooperate by putting souls in at just the right moments. 3) Gods are not cooperating, so now we are embarked on creating a new form of life without souls. All of these are provocative. Even if you are not religious, the first alternative has some uneasy implications when we get around to making silicon-based robots that look and act just like humans. What will we have that they won't?

Up to now, we have been trying to hold the new meanings of life together, rearranging a little, re-interpreting a little, redefining a little. All of these new angles came from things we've recently discovered about the kind of life that already exists on this planet. Finally, we need to consider the wild frontier that comes from understanding a naturally occurring phenomenon well enough to make artificial versions of it. We learned something interesting from our experience with artificial intelligence research over the last 40-some years. The discipline began as an attempt to better understand human intelligence. With the dawn of the computer, we had a way to artificially model human intelligence. There's nothing quite like building a working version of something as a means to understanding it. Early researchers felt somewhat compelled to build programs that emulated the way humans actually performed intelligent tasks. We quickly learned that this was hard. We could, however, easily build programs that performed much *better* than humans on certain intelligent tasks if we did it in a different way – programs that were arguably *more* intelligent than humans. This caused the discipline to split into two branches: an academic one whose goal was to model *human* intelligence in order to better understand it, and an engineering one whose goal was to build intelligent machines and devices to solve intelligent problems in the most effective way possible. The second discipline was freed from the constraint of emulating human intelligence. Intelligence became a more general concept. You could improve it by adding on new aspects and jettisoning others to best fit the problem you were trying to solve. Now this is happening with life.

One Small Step for Bacteria-kind

In May of 2010, scientists at the J. Craig Venter Institute announced that they had created the first self-reproducing bacterium controlled entirely by a chemically synthesized genome. The scene was very different than that from the lab of Victor Frankenstein. Instead of a solitary scientist in a darkened room throwing the switch on a secret experiment assembled surreptitiously through the black arts of metaphysics, this was the result of some 46 researchers, leveraging the prior, published work of others, over the course of 15 years. There were reused body parts involved, but no secret sauce. When you put the parts together properly, life happens all by itself. The potential is already in the parts. There was no danger of the man-made organism rising up and terrorizing the local villagers, but it was a designer bacterium modeled after *Mycoplasma mycoides*, a subspecies of which causes lung disease in cows and goats.

Victor Frankenstein's result was much more dramatic because he went for a whole, synthetic human right from the start. The Venter scientists were less ambitious, going for the closest practical thing to the minimum life. This is what real scientists do, of course: solve the most basic version of the problem first, then build their way up. The project was originally aimed at something even more minimal. Dubbed the *Minimum Genome Project*, it began in 1995 to synthesize the least genome that could support life at all. This was both a practical consideration and an attempt, by building a working version of them, to understand what the minimum set of genes for life is. The smallest known genome at the time that could be cultured in the lab belonged to *Mycoplasma genitalium*, coming in at 475 genes. By deleting individual genes one by one, it was determined that only 382 of these are necessary for the organism to survive. So this minimal set was the original target. It was to be called *Mycoplasma laboratorium*, in honor of its unique origin. But *M. genitalium* is relatively slow growing, so the project switched targets to two larger, but faster reproducing species, *M. mycoides* and *M. capricolum*.

The event wasn't quite like the original abiogenesis, both because we started with a much more evolved, modern design for an organism, and because only one of three elements of life was synthesized from scratch from its constituent chemicals: the genome. This was a custom variation of the *M. mycoides* genome, called *M. mycoides JCVI-syn1.0*. The cytoplasm and cell membrane from an existing M. capricolum (with its native genome removed) was used for the other two parts. Still, this would have been like Victor Frankenstein synthesizing an entirely new brain from chemicals, then transplanting it into the assembled body parts (minus the original brain). The genome contains all of the instructions for what proteins and RNAs are expressed in an *M. mycoides JCVI-syn1.0* cytoplasm, and how to regenerate and maintain the cell membrane. Once the new, hybrid organism reproduced enough times, the M. capricolum version of the cytoplasm had become recycled to match that of the new genome. At that point you have two of the elements synthesized.

Eventually, when the lipid membrane has been completely recycled, all three elements are new. The body of an M. capricolum was essentially used as a factory to build the body of an M. mycoides JCVI-syn1.0.

There were other body parts involved as well. Biotech has long used single-celled organisms as factories to produce proteins and other useful molecules, and to assemble DNA sequences. Nature already provides us with some wonderful micro-machinery for doing these things, so why not reuse it? The new genome was first designed by computer, then created in pieces by chemically synthesizing its 1078 DNA segments (each 1080 letters long). The whole genome was then stitched together from progressively larger sub-pieces of the DNA by using the DNA replication machinery in a yeast cell. The creators even signed their work by embedding a DNA watermark in the genome, so that all future progeny of the synthetic organism can be identified. Because life is so tolerant of DNA errors and large sequences of non-coding DNA, there are plenty of places in the genome where you can use the four DNA letters (A, C, G, T) to spell things – the functional equivalent of "Kilroy was here." The researchers first encoded a custom translation table in the genome from sequences of the four DNA letters to the 26-letter Latin alphabet, then used the new coding scheme to spell out the names of the 46 contributors, a web URL that can be used to verify the watermark, and some choice quotations from James Joyce, Robert Oppenheimer, and Richard Feynman (the one we used above). A new perspective on writing in molecules.

Biotech has been creating novel organisms for some time by adding and subtracting genes from existing organisms. What's new here is that the (approximate) functionality of an existing organism has been reproduced from an entirely synthetic genome. It is the manufacturing process that is novel. It demonstrates that genomes are genomes, whether nature makes them or we make them. It may seem like a long way from here to artificial humans, but remember, almost all of the variety in Earthly life is encoded in genomes, including how you grow up a whole human from a single cell. We already know the entire code for a human. Synthesizing a custom version of that genome with a process like that used for M. mycoides JCVI-syn1.0 is now largely a matter of scale. There are only two basic varieties of cell body: prokaryote and eukaryote. It will be no small engineering feat to synthesize a eukaryote cell body from scratch, but once that is done, all you have to do is add your custom-made genome and the rest takes care of itself.

So this is both a watershed event, and a non-event. It is significant because it is the first of something. But it is something that we expected to happen sooner or later. The principles were already well understood. We are now working on the engineering minutia. As usually happens, the first of something takes a long time (15 years) and a lot of money ($40 million). This happened with the first sequencing of the human genome. Then technology improves, and manufacturing efficiencies kick in, and the time and cost fall exponentially. Just four years later, a group of researchers composed largely of undergraduates has just synthesized a drastically redesigned version of a whole eukaryote chromosome. When transplanted back into the yeast from which it was modeled, it behaves just like a normal yeast chromosome. Now the goal is to synthesize a remodeled version of *all* of the yeast's chromosomes by farming out the project for each chromosome in parallel to a different university lab where again the synthesis will be done largely by undergraduates. Crowdsourcing.

Life Plus and Life Minus

If we view artificial life as an engineering discipline, similar to what happened with artificial intelligence, we are freed from the constraints of its current biological forms. We don't need to build complete organisms, or limit ourselves to what current organisms can do. We can make things that are almost life, or life-inspired, and things that are more than life. At some point it becomes irrelevant whether this is still life properly so called, as long as we make something useful. The earliest successful industrial applications of artificial intelligence, for instance, were the ones that specialized in some very narrow problem domain, and threw away every other aspect of a human that didn't bear on the task at hand. If you need to spot-weld car chassis on an assembly line, for instance, you don't need a head, or legs, or the ability to talk. You only need one very specialized arm. If you need to spot investment trends, or to make medical diagnosis, you don't even need a body. An intelligent computer program will do – one that can do one and only one thing as well or better than humans can.

One big difference from artificial intelligence is that artificial life almost always gets to a solution quicker by slightly modifying some existing instance of natural life than by trying to build something from scratch. That's why M. mycoides JCVI-syn1.0 is such a yawner from an engineering perspective. All that work and it does the same thing! At some point in the future, when artificial intelligence, robotics, and artificial life all converge, and we are able to manufacture silicon-based versions of familiar organisms, we will have some new conceptual problems concerning the meaning of life (both logical and ethical). But for now, the existing machinery of natural life is so modular and so robust that there's no practical point to making artificial life out of anything but carbon-based chemistry, and to do that by slightly modifying an existing life – and to use the machinery of existing life for the fabrication as well.

Some of the most innovative things are now being created down at the nanometer scale – at the level of bacteria, the smallest of the whole life forms, and even smaller bio-things such as viruses, and vesicles and proteins, which are sub-life components. A problem we have, as large beings fighting diseases at the cellular level such as cancer, is how to get drugs delivered to just the cells that are affected by the problem without killing good cells. Our current chemotherapy is a system-wide, carpet-bombing approach that targets all fast-growing cells. This gets the cancer cells, but it also gets the skin cells, and the hair follicle cells, and the digestive epithelial cells. Our immune system, on the other hand, takes the battle to the enemy at its own scale – cell to cell. We will soon be able to emulate this more targeted warfare with some new quasi-life forms that possess some of the features of immune cells, viruses, and vesicles. These nanoscale drug delivery vehicles need some sort of container to hold the pharmaceutical molecule, some way to recognize and navigate to the target cell, and some way to get inside the cell to deliver the payload. Vesicles (recall the spherical lipid bilayer) deliver cell payloads by merging with the cell wall. Viruses do it with special protein needles. All three bio-agents navigate and recognize with surface proteins that bind to surface proteins on their targets. Researchers simply design and create new micro-bio-agents with the best set of these mechanisms for the job. New forms of artificial life? Almost.

Another nanoscale bio-thing popping up in labs all over the world is the bio-computer. It was bound to happen. You can make small circuits and computing devices out of all kinds of materials, but if you make them out of DNA and RNA and proteins, you can embed little computers in the body to monitor things, calculate thresholds, signal the presence of things, take corrective action. If we get really good at this, perhaps we will be able to make our future intelligent robots out of squishy carbon instead of stiff silicon, eliminating some of their glaring differences from natural humans.

A lot of the work in artificial sub-life is directed at designing better proteins and other biomolecules, and better nanostructures such as fibers and tubes that can be used in industrial applications outside of life. These are often inspired by naturally occurring biological structures of existing organisms. Nature has had 3.5 billion years of evolution to experiment with all kinds of materials and structures, so it pays to copy some of these designs, perhaps with some slight tweaking. There is a relatively new field of research called *biomimetics* that takes exactly this approach. Find an existing, near optimal solution in nature and repurpose it for industrial uses. Spider silk, for instance, is better than Kevlar. The mimicking also goes the other way. One of the problems facing regenerative medicine in the growth of new tissue and organs from stem cells is how to emulate the fine vascular structure of blood vessels. More than one lab has approached the problem by noticing the similarity of this structure to cotton candy. Cotton candy machines have actually been used to create the original scaffolding, either by spinning real candy from sugar, coating it with a liquid polymer, then dissolving the sugar, or by modifying the machine to spin protein fibers to begin with. Industry mimics life; life mimics industry.

Most of the artificial life action so far, though, has been aimed at creating artificial bacteria. As we've seen before, bacteria occupy a unique position in the Earthly ecosphere. They are the biological substrate that mediates between life and inorganic chemistry. If you need to make some molecule X out of some other molecule Y, chances are one or more bacterial species somewhere already does that. Humans have been using bacteria as chemical factories since the dawn of agriculture. It is estimated that as much as a third of the human food supply is the result of fermentation by either bacteria or fungi – things you already know about, like beer, and wine, and pickles, and cheese, and yogurt, and things you probably don't know about, like chocolate and coffee. Humans originally recruited these symbionts without realizing they were there. Pickling, for instance, was first discovered, like most fermentation processes, as a preservative. If you put cucumbers in an appropriately briny solution at the right temperature and humidity, they not only become pleasantly sour, but they don't spoil. What's really going on is that this environment selects for a particular species of bacteria that produces lactic acid as a by-product. That's the sour part. The preservative part is due to the fact that this particular species does so well in the briny environment that they outcompete all other species of microbes that would otherwise be eating your cucumbers. That's what "spoiling" is – micro-creatures eating your food before you do. Your favored bacterial partners are also eating your cucumber, but much more slowly, leaving plenty for you.

In the modern era, we began using bacteria and fungi to manufacture antibiotics, once it was discovered that they make these toxins naturally to kill each other. Now we swap their genes and redesign their metabolic pathways to create factories for just about anything we want. They can be reengineered to produce biofuels from organic and inorganic compounds, to produce biodegradable plastics to replace those derived from petroleum, to produce organic textile fibers for carpets. They can be redesigned to purify wastewater and generate electricity as a by-product. They can be deployed to clean up hazardous waste sites and ecological contaminations. They can be redeveloped to act as biosensors, emitting a signal molecule or a light wavelength in the presence of environmental phenomena such as humidity, organic pollutants, viruses, other bacteria, hormones, drugs, DNA sequences, heavy metals, or toxins.

The early days of artificial microbe design have resembled the early days of software design. First came the scientific principles that enabled individual practitioners to create novel solutions. Everyone then set off in parallel to roll their own, often duplicating efforts and building custom solutions that could not be easily reused for other applications. When software engineering grew up, it changed into a discipline that emphasized standardized parts, and reusable architectures, so that developers could cobble together new solutions largely by reusing existing parts from other developers. Synthetic bioengineering is now beginning to make this same transformation, by building catalogues and registries of standardized bio-parts and bio-architectures, so that labs can collaborate and easily reuse each other's work. One of the trends that emerged from reusable software was the notion of *refactoring*, revisiting the designs of existing components with hindsight and redesigning them to be simpler, more streamlined, and more modular. The best design is often not apparent until you've built the first, non-optimal version of something. The natural model organisms for microbial systems, such as *E. coli* for bacteria and *S. cerevisiae* for yeast, are now being considered for complete resynthesis so that they are simpler and easier to understand as standard platforms for creating new species. Here we have a case where the bumbling, blind process of evolution has created inelegant things that manage to work, and now humans are retroactively taking on the erstwhile divine role of the intelligent designer.

Microbes are easy to work with because they are relatively simple, they don't cost much to maintain, they multiply as quickly as every 20 minutes, and nobody complains about whether you are abusing them, or enslaving them, or harvesting them (the euphemism for killing them to extract something of interest). Artificial life at the *multi*cellular level has mostly been about plants – the GMOs we looked at earlier. For animals, the action has been primarily in what are called knockout mice. We don't really need a better mouse, for ordinary mouse purposes. But since this is the model organism for mammals, and has enough common genetic and physiological traits to be used as an invasive experimental model for humans, endless versions of novel mice have been created (and some even patented) that are missing one or more of some genes that they share across the mammalian line (particularly with us). This is the sense of 'knockout.' The target gene has been removed from the mouse genome to see what its system-wide role is in mouse physiology. Whatever the new mouse is missing at the phenotypic level, or whatever physiological process no longer works the same, indicates what the erstwhile gene should have been doing.

Perhaps the most provocative form of artificial animal life so far is not a whole life at all, but one of its living parts. Stem cells have mostly been used to grow tissue for regenerative medicine – to replace parts that are broken or damaged or diseased. But you could also grow synthetic tissue in order to eat it. It is one of the inescapable facts of life on Earth that life eats life to survive. We all do it, every species. We all need the same basic organic building blocks of sugars, proteins, amino acids, and nucleic acids. Most of our bodies cannot synthesize everything we need, so we have to get these nutrients from other organisms. It's a kind of theft, but that's how the great cycle of life works. The bacteria are the only ones that can synthesize organic stuff from inorganic stuff, so they provide the ultimate input for the cycle, but they also show up to take their turn at the organic table, slowly eating away at big organisms once they die and their immune systems shut down.

Vegetarians decry the practice of humans eating other animals, but are OK with stealing the lives of plants. Plants, though complex life forms, are thought to be in a different religious or moral category. They are our food; the animals are our brothers. Some of this has to do with the fact that plants, lacking nervous systems, feel no pain. We have the luxury as modern humans of being vegetarian because we have completely reengineered our plant cousins to contain enough nutrients to sustain a metabolism that was evolved around eating the higher energy and protein content of meat. It was our transition to carnivorous hunters that supported the higher energy demands of our bigger brains. We are adapted to eat cooked meat, and have taste and olfactory sensors that incline us toward that practice. The lean muscle tissue of other animals is good people food. When you steal life from another animal, you get all 20 amino acids in every bite. When you steal life from plants, you have to kill a good variety of them because any one has only a few of the proteins you need. It can be done, but you have to be diligent.

Well, what happens when you don't have to steal life from another animal to get your most efficient food? We don't need the whole cow; we just need the lean muscle. So why not synthetically grow our "beef" from bovine stem cells? No actual animals were harmed in the making of this movie. No one felt pain, or lost a parent or a child or a sibling. No feed lots, no factory farms. The individual *myocyte* cells in our synthetic beef are alive, until cooked, but they never reach the threshold of a whole animal. There is no nervous system. We are merely stealing life from individual cells, something we are already comfortable doing with plants and fungi. So meat, as a food category, can be divorced from its live animal origins. Synthetic meat would be even less alive than potatoes or mushrooms. Vegetarians often surmise that modern humans would lose their taste for meat if they could view its source at the slaughterhouse instead of in the supermarket aisles – see how it is essentially different than potatoes or mushrooms. Well, now it doesn't have to be essentially different. The upside, nutritionally, is that this is real beef, not some GMO derivative. This is exactly the thing we evolved to metabolize "naturally." This has already been done in The Netherlands, and in May of 2013, the first artificial beef burger was cooked and eaten at a news conference in London. One of the tasting food critics described the experience this way:

> There is really a bite to it, there is quite some flavour with the browning. I know there is no fat in it so I didn't really know how juicy it would be, but there is quite some intense taste; it's close to meat, it's not that juicy, but the consistency is perfect. This is meat to me... It's really something to bite on and I think the look is quite similar.

Silicon Persons

Because it is so easy to create artificial life out of natural life, the first synthetic creatures out of the chute don't really challenge our existing meaning of life all that much. They are still contained metabolisms with heritable reproduction. They even conveniently use the same biochemistry, even though we don't regard that as essential. But if we fast-forward to some point in the future where we can make our current silicon-based robots look and behave exactly like humans, we will have something that doesn't qualify as life by *any* of our current criteria – not just because of what they are made of, but because they have no need for a continuing metabolism, and they don't need to reproduce themselves. Yet it will be very hard not to classify them as intelligent persons, the highest, most complex form of life that we know. This will rend the fabric of the meaning of life like nothing else we can imagine. We aren't tempted to think of current robots as alive because we are not remotely fooled into thinking they are humans. But when we can't tell the difference, when they speak and act and cry out in pain and empathize with us like real persons, we will think they are alive, and we will treat them like real persons. We will respect them, we will behave ethically toward them, and we will suppose that they have the same civil rights as us.

The conceptual and ethical dilemmas will occur when we first discover they are actually robots. Does this suddenly make a difference? Do what they are made of, where they came from, and their childfree lifestyle matter? Because of our inbuilt empathy for the emotions of others, we already extend a certain moral and filial standing to the charismatic animals that are our pets. They don't qualify for everything because they are "just animals", but we take care of them like children, we include them in our wills, we mourn their deaths. We don't "treat them like animals" the way we would cockroaches or snakes. Yet we know they don't have full-fledged souls like humans. We know this by their appearance and behavior. We have a tougher time with chimps and bonobos because their appearance and behavior is much closer. What would we think of *Homo erectus*, or *Homo neanderthalensis*? Very close.

Throughout our history, we have shown the capacity to discriminate against certain classes of real humans because our social customs lead us to classify them as not quite human. We have exploited the members of competing tribes, enslaved people from different races, ostracized and persecuted people with different religions. We could only tell they were subhuman by their stereotypical appearance and social context. When we encountered uncharacteristic variants of them who managed to make a credible living inside our societies, there was always that awkward moment when we first realized that this person was not really the full person they had led us to believe they were. Some of us felt deceived, and mentally disassociated ourselves from them. Some of us learned a valuable lesson about the arbitrariness of our categories and welcomed them into our societies. This is what it will be like when we first unmask our perfectly android robot friends. Will we suddenly see them as non-persons, or will we give up the notion of the intangible essence that they lack?

If you are religious, of course, you will have to decide whether you were wrong in previously thinking they had a soul. But even if you aren't religious, you will have to consider whether you were wrong about their civil rights, their moral and legal standing, whether you can treat them as property, whether they really feel anything even though they show visible emotions. Philosophers of the past have used examples of imagined automata to mark the distinction between persons and soulless machines. Both are mechanical in their own way, but the former are machines plus souls; the latter, just machines. Descartes thought this same distinction extended to animals – they didn't have souls because they were essentially mechanical automata. The soulless human-like automata *had* to be imagined because the closest analogues of the time were things like clocks and steam engines. Not very impressive. No one expects these things to feel emotions. But this wasn't a fair thought experiment. It is now easier for us to imagine robots getting perfectly android-like, so the better thought experiment is the scenario of a light-skinned African American being mistaken for a Boston Brahman in the 19th century, or a Jew being mistaken for a Christian. Could you really bring yourself to reconsider their moral or civil standing?

We are still a long way from silicon persons, but we are close enough to see how our current, very fluid meanings of life might have to change in more fundamental ways to accommodate them. Someday we will have to answer these questions in a practical way. From our present perspective, we can already see that the most significant meanings of life for humans involve things like morality, community, rights, obligations, and dignity – things not essentially connected to biology except for how they came to be in the first place. If there are other ways to get there, we may have to generalize our current meaning of life beyond the green, squishy stuff.